D1624758

HYDROGEN BONDING BY C—H GROUPS

Though armed with the Watson sisters' indispensable guide, I would not have been able to figure out how to approach the site the only way one can nowadays—over logging roads cut deep into the woods and brick-red South Carolina clay—had I not been blessed with the good luck of meeting a man who knows the terrain well. Eliott Werts, now retired, lives at the edge of the bright green tract, which, in times past, he owned, then sold to the forest products company that operates it; and it was at his house that I met him with my request for help in finding my old uncle's resting place.

Obligingly, Mr. Werts dropped his yard work and took me down a path, over a fence, and into the timbered wilderness, telling stories about the land as we went. Time was, he said, when the hills from the Saluda inland up to the site of his present house—and far beyond, almost as far as the nearby town of Ninety Six—lay bare to the sunshine, and full of cotton. But that hadn't been so for a long time. A late aunt, who had lived on Samuel's land years before, once told Mr. Werts she recalled seasons when an unobstructed view lay open across the fields, beyond a range of beautiful, low hills, to the distant horizon. She showed her nephew the spot on which Samuel Mays's spacious manor had stood, and his various outbuildings, and the low hill on which Samuel had buried his slaves, and where his widow and children buried him; and all these things, Mr. Werts showed me.

I went to the place again on a chilly, brilliant winter afternoon a couple of years after the first visit. On this occasion, my guide was Dr. Samuel H. deMent, a young pathologist at the Self Memorial Hospital in Greenwood, South Carolina, just a few miles past the oddly named town of Ninety Six from the pine stand we walked in. Over the eight seasons since I had last been there, the softwood trees had undergone startling growth, occluding what little of the rough landscape we could then see from the company road, leaving open to view now only the clear, cold sky and the muddy lane ahead, puddles reflecting the dense verdure and afternoon radiance, and the shadows of empty deer-hunting stands, on their tall scaffoldings nestled into the trees on the road's margin along the way.

The doctor is a man with a special claim to this remote spot. His interest in Southern memory had first been sparked in the usual way, with quickened curiosity about his French Protestant ancestors, who had been expelled from their homeland and found their way to the South before the Revolution. But like most serious genealogists, he stayed fixed on the dry business of constructing family history for only a while before wanting to know more, and to remember more broadly. Dr. deMent joined the Sons of the American Revolution in Tennessee in 1991, and shortly found himself at the head of the committee of his state society responsible for tracking down and marking the forgotten graves of veterans of the Revolution. Among the first he discovered and marked with a plaque was that of Samuel *Mayes*—possibly a shoot off the family's branch that also produced Edward Mayes of Mississippi—in the Presbyterian churchyard at Mount Pleasant, Tennessee.

Soon after moving to his present position in South Carolina, he discovered the name of Samuel Mays, my near kinsman, in the Watson sisters' book, and resolved to find his gravesite, which he did (as I did later) with Eliott Werts's help, and in 1994 saw to the erection of the small, sturdy headstone that stands there now. Of the eighteen Revolutionary warriors whose burial places he had found, Dr. deMent reported to his chapter of the SAR, "the two Samuel Mays(es) will always be special. Unfortunately, I cannot claim kinship to either Patriot, nor can I identify a direct descendant to share their amazing stories. However, I have found that wherever I am, people have been helpful with the grave markings and appreciate the need to preserve and maintain the few remnants of American Revolutionary War history available for our future generations to observe and respect . . . before these sites disappear in time."

As it happens, Dr. deMent is probably one of the last people to know this spot, which has already virtually disappeared from sight and memory, and one of the few who will ever visit and honor it. The heavy stone he bore into the pine forest and set down is a sentinel by the grave of a notable soldier and statesman of my blood, and for his putting it there, I will be forever in Dr. deMent's debt. But if his commemoration of the two Samuels remains special to him, my trek back into the bush that winter's day with this man *who remembers* will be special to me until my last day, and among the most poignant souvenirs among all my voyagings in search of my kin.

The plain, low slab, inscribed with a cross, the words *Samuel Mays*

Outlet mall, Virginia

Pvt SC Militia Rev War, and his dates, is aglow with wintry afternoon light in my mind as I write these words—a memorial to a virtuous man, but also a fitting tribute to Samuel Mays's ancient idea of the land that he made his own. It stands far away from the nearest road, at the center of his five thousand acres, tended by Samuel to guarantee his authority and identity as an American man of the soil; but the land long ago passed from his family's hands into others. Kept away for more than a century after his death, the usual wild, sudden surge of trees and brambles over unkempt land devoured it during the Great Depression. And finally came its planting with pine seedlings— the conclusion of the centuries-old struggle by the Southern farmers and graziers on that ground, and by imperial landowners such as Samuel, to keep clear the cropland and pastures. I have known the grace of cleared cotton fields, their sunny calm and twilit splendor at Spring Ridge—the plain grace singing from the heart of all the agrarian visions of Southern renaissance that sing no more. And I have known the abandonment of our Southern fields, their swiftly rising horizons of trees, their vanishing under malls or forests, and felt it as the loss of more than fields—of a way of being as well, rich in the gravity and joy born in the South from the marriage of Rome and our native soil. I may never revisit the grave of Samuel Mays. But the simple stone I left in the gathering twilight will glow in memory until my own dusk comes, and all the memories of ancestors and the land I cherish are swallowed up by night.

HYDROGEN BONDING
BY C—H GROUPS

R. D. GREEN
Department of Chemistry, University of Saskatchewan

A HALSTED PRESS BOOK

JOHN WILEY & SONS
New York – Toronto

Library
I.U.P.
Indiana, Pa.

547.01   G825h
C.1

© R. D. Green 1974

All rights reserved. No part of this publication
may be reproduced or transmitted, in any form or
by any means, without permission.

First published in the United Kingdom 1974 by
THE MACMILLAN PRESS LTD
London and Basingstoke

Published in the U.S.A. and Canada by
HALSTED PRESS
a Division of John Wiley & Sons, Inc., New York

Library of Congress Cataloging in Publication Data

Green, Robert D
    Hydrogen bonding by C—H groups.

    "A Halsted Press book."
    1. Hydrogen bonding. 2. Chemistry, Physical
organic. I. Title.
QD461.G75   1974   547'.01   74-11310
ISBN 0-470-32478-3

Printed in Great Britain by
Thomson Litho Ltd.
East Kilbride, Scotland.

Preface

The hydrogen bond is a well documented phenomenon, having been found and characterised in a very wide variety of systems of chemical interest. The literature of hydrogen bonds can only be described as immense, as witnessed by the comprehensive 1960 monograph *The Hydrogen Bond* by Pimentel and McClellan which lists over 2000 references up to about 1956. There has been no abatement of interest in the subject since that time and the book *Hydrogen Bonding* by Vinogradov and Linnell, published in 1971, makes no attempt to collect all the relevant literature of the intervening years.

It has always been recognised that the most important classes of hydrogen bonds are those involving O–H and N–H groups. Hydrogen bonded to other atoms may also form hydrogen bonds, but these are, as a rule, weaker and less frequently observed. Consequently, Pimentel and McClellan's book [309] deals with the subject of C–H hydrogen bonds in four pages, while that of Vinogradov and Linnell [389] devotes a few scattered sentences to the subject. Hence, in view of the continuing interest in C–H hydrogen bonding, which has already resulted in more that 300 papers dealing with them, it was felt that there was a need to have the knowledge of the subject collected together so as to facilitate the evaluation of our present knowledge and the formulation of fruitful approaches to further study of the phenomenon.

Chapters 3, 4 and 5, which form the bulk of this book, contain collections of the available information concerning the hydrogen bonding properties of sp^3, sp^2 and sp C–H bonds respectively. Wherever possible, the data have been collected in tabular form and reduced to a common basis. For example, NMR chemical shift differences between solvents are all listed in parts per million (ppm) and the two appropriate solvents are given.

No attempt has been made to provide a complete discussion of the various methods of detection and characterisation of hydrogen bonds; such a discussion is to be found in Pimentel and McClellan's excellent book [309]. However, a short review of those methods which have been of greatest use in examining C–H hydrogen bonds comprises Chapter 2.

The bibliography is as complete as it could reasonably be made, through 1972. Some papers whose mention of C–H hydrogen bonds is peripheral and add nothing to our understanding of them, have been omitted. If there be any further omissions, these were not intended.

The nomenclature of chemical compounds is throughout as nearly systematic in line with IUPAC recommendations as possible (see *Nomenclature of Organic Chemistry*, Butterworth, London, 1969). In a few cases, non-systematic names have been used in an attempt to make a point clearer: e.g. triethoxyphosphine oxide, rather than triethyl phosphate for $(C_2H_5O)_3PO$. Data have been converted into SI units wherever they are given in the literature in other units.

Thanks are due to the University of Malaya for its support, particularly for my extensive use of the library facilities and for some assistance with the typing. Especially appreciated was the co-operation and assistance of Mrs Agnes How who sought out and obtained scores of articles from various obscure sources.

September 1973 R. D. GREEN

SEEKING THE OLD SOUTH

House porch, Henderson

B Y THE TIME I CROSSED THE SALUDA RIVER ONTO THE FARM-lands of Dorcas and Mattox Mays's children, I had tra-versed the geography of almost two hundred years of my family's history, but still not come to the popular modern imagination's Old South. Cotton had hardly gained a toehold in the Southern United States by 1816, when Samuel Mays died; the export field crops were still what they had always been, tobacco and hemp and, in the low-lying districts, rice. In some states, the plant destined to become almost synonymous with the Old South was at first actively resisted as an agricultural novelty. The Virginia legisla-ture did not get around to encouraging cotton-growing until 1827, and it took another decade after that for the Old Dominion's first spinning and weaving operations to begin production. There were, of course, many black slaves in the South in the early nineteenth cen-tury, but only a fraction of the number who would be shackled to the land after the advent of large-scale cotton cultivation. And as for storybook belles—they, too, were products of the culture brought by cotton to the South, and of the wealthy, increasingly rigid agrarian regionalism that followed the South's degradation to a one-crop economy.

The usual date given by historians for the emergence of the Old South is 1830. It lasted for only some thirty years in South Carolina, during which time, the historian Louis Wright has said, the state "exemplified the ultimate in social and political perversity, the wastage of brilliant minds fighting for futile (and often iniquitous) causes, and the enthronement of that worst of the Seven Deadly Sins, Pride. Even more terrible, South Carolina led the whole of the South to ruin."

But while Wright's reading of the cotton culture in which the sensibilities of my antebellum ancestors were formed is, I under-stand, academic orthodoxy nowadays, it was not the idea I grew up with. That older idea, I learned—along with millions of other South-erners my age—from *Gone With the Wind*.

When I was still a child, in the late 1940s, viewing *Gone With the*

Contents

Wind was a dress-up event, like church and shopping and funerals. For my grandmother Erin and Aunt Vandalia, it meant hat and gloves, fox stoles and pearls; for the boys and menfolk among my Mays kin, suits and white shirts and ties. The famous film was big, like Cecil B. deMille's *Samson and Delilah*, requiring a theater with a large screen; and, because huge crowds attended, its presentation required a large picture palace, of the kind still, at that time, found only in urban downtowns. Hence, going to see it, like serious shopping, involved a "trip to town"—a phrase that rings even now in my mind with a gravity of occasion soon to be deflated, once the interstate highway had abolished the distance between the Louisiana village of Greenwood and the city of Shreveport, and the malls and suburbs had emptied the urban core of its old importance.

But *Gone With the Wind* was big, at least for white Southerners, in another sense. I recall the privileged air of ritual in seeing it, the almost sacramental quality of being there—as though watching this Dixie *Götterdämmerung* unfold, to the accompaniment of Max Steiner's grandiose musical score, was to be reborn in the imaginative universe conjured in the opening titles on the screen. "There was a land of Cavaliers and Cotton Fields called the Old South," they told us. "Here in this pretty world Gallantry took its last bow. Here was the last ever to be seen of Knights and their Ladies Fair, of Master and of Slave. Look for it only in books for it is no more than a dream remembered—a Civilization gone with the wind." After these words scrolling past glowed a stately sequence of watercolored vignettes of sun-washed cotton fields and docile darkies and toiling mules—a frieze of images summarizing in moments an entire library of lyrical notions embedded in the phrase *Old South*.

Seeing the film in my single-digit years, I found it foreign. I had never thought of myself as a Southerner—at least with the peculiar connotations I was later to learn about—though I did know, of course, that I lived somewhere called the South, and had spent my earliest years on a cotton plantation. But neither region nor childhood circumstances, nor hardly anyone I had known, bore much resemblance to the places in *Gone With the Wind*. Only the blacks— ample and devoted Mammy, silly Prissy, and Big Sam—were at all familiar. The sinister role in the film was confined to the Yankee bloodsuckers and carpetbaggers, and the white trash I had been reared to loathe. But in the culture of my young years, however, whatever affection the white people around me may have felt for black servants or laborers was tinged with the usual, universal racist fears of the Other—the unspoken anxiety over the sexual violation of "our" women by "their" men, apprehension about "uppity niggers," the menacing thought that someday "we" would be overwhelmed by "them." If Scarlett's Mammy seemed more real to me than anybody else in the film, it was because my formative childhood experience of blacks had been devoid of anything remotely resembling menace. My own mammy, Essie De, was the incarnation of loving attentiveness, unstinting in generosity, even when such generosity was not demanded. As the white people around me—my father, and mother, and sisters—came and went, then died or went off altogether, Essie De was always there. When my mother and I abruptly left Spring Ridge in the autumn of 1947, and moved to Shreveport, Essie De did not have to leave the countryside she'd lived in all her days, but she came anyway, to care for me then, and in the years after the death of my mother. Like Mammy in the movie, my own mammy had been as subject to imperfection as the rest of us, but always stationed firmly, freely, in the middle of my existence.

If I had known somebody like Mammy, I had never met any white people like Rhett Butler and Scarlett O'Hara, and the boyish swains who courted Miss Scarlett, or like Mr. O'Hara, at least in his antebellum period. The postbellum, insane master of Tara seemed plausible, perhaps because the upcountry of my childhood was dotted with mad folk, white and black, shuttered in their shacks or mansions, and known to us in rumors about places we children were not to wander. But I had never encountered a person with Mr. O'Hara's Irish accent, and only one or two with his Catholic religion. Oddest of all, however, was the strange way all the Southerners in the movie acted, as though afflicted with a disorder that made every stance a contrived posture, every scene an allegorical tableau. But I decided that was just the way people in movies about the South were *supposed* to behave.

I saw the film more than once as a child and early teenager, then saw it no more until the middle years of the 1960s, when I was a newcomer to graduate study in Rochester. I probably would not

List of Symbols

A–B   a 'normal' covalent type chemical bond

A···B   a weaker bond (e.g. a hydrogen bond)

$K_\phi = \dfrac{\{AB\}}{\{A\}\{B\}}$ equilibrium constant for formation of a complex, where $\{X\}$ is the activity of species X

$\Delta_\phi H, \Delta_\phi S, \Delta_\phi G$ enthalpy (entropy, Gibbs free energy) change upon complex formation

$\Delta_m H$ enthalpy change of mixing

$\Delta_l^v H$ enthalpy change of vaporisation

B second virial coefficient

ν wavenumber (in cm^{-1}) of a vibrational mode

$w_{1/2}$ full width at half height of a vibrational band in the infrared or Raman spectrum

A   intensity of infrared absorption band

Δ_c chemical shift difference between associated and non-associated molecules

Δ_s chemical shift difference in two solvents

have done so then, had I not been invited to a screening by a fellow Southerner in a *Beowulf* seminar.

We had become acquainted through a shared enthusiasm for Anglo-Saxon literature. Our friendship began, however, when we discovered both of us had been worrying about a common matter: the assumption by other students in our seminar that she and I were *stupid*, probably racist, and otherwise undesirable. This was the era of civil rights, after all; and Northern classmates, in our experience, tended to greet any white person from the South with a certain hostility in their back pockets, if only because we all sounded to them like Strom Thurmond and George Wallace. (There would have been no point in arguing that there are hundreds of "Southern accents," differing among regions, classes, even families.)

I was somewhat accustomed to feeling alien, due, at least in part, to the tightening grasp of depression. My friend was not. The daughter of a distinguished tidewater Virginia family, she had grown up in comfortable suburban circumstances, secure in the knowledge of who and what she was, and of what her family had been. She knew, as they say, *her place.* Like other Southern women of similar breeding, she had done the circuit of private schools in the South and abroad, and graduated from a respectable Southern university. Unlike many another beautiful tidewater sister, however, she had emerged from this traditional grooming as that most remarkable entity, a belle with brains—charming and elegant enough to become a Southern senator's wife, perfectly at ease in the genteel pleasures and hypocrisies of her class, yet brilliant enough to run intellectual circles around us all. She was also furious. For, unlike me, she had decided *not* to tighten up her drawl to stop the snickering that flickered around the seminar table when she read aloud her Old English papers. Nor was my friend prepared to tone down the distinctive, faintly snobbish, but quite lovely Southern hauteur she'd only recently discovered there was something wrong with. (I had been snarled at quite enough by Northern New Left women in my classes, when starting to help one with her coat, or stepping back to allow her through a door first; and had stopped doing so.)

Going to see the great Southern cult classic, and telling everybody she planned to do so because she just *adored* the film, marked her closest approach to revenge during our seminar days. She actually hated *Gone With the Wind*, I learned beforehand, for purveying to Americans at large the absurd version of the South that had now come back to afflict her—a vision of our region as a sort of hothouse of exquisite, ignorant gentility, banal manliness, flirtatious and frivolous femininity, all with an underlying cultural agenda of racism. I shared none of this animus, and had few feelings of any sort about the film, but was pleased by her invitation, so went.

After staggering out of this very long movie she and I had been raised on, and collapsing in a nearby bar, we began to talk, and talked for hours about what we'd seen. We had both gone in expecting one thing—the potted, brittle Old South of our childhood thinking—and we had gotten another. If *Gone With the Wind* was supposed to a sort of *Gloria* to the Old South—as both she and I knew it to be, at least in the minds of myriad Southerners—then *where had the Old South gone*? The antebellum rural world is glimpsed for only a few minutes at the very start of the film, in the scenes at Tara, where we meet the flirting, scheming Scarlett O'Hara, and at the Wilkeses' palatial Twelve Oaks plantation—then disappears. Its remaining hours, set in the years of Civil War, Reconstruction, and Southern recovery, could be read quite coherently, to our surprise, as a devastating dismissal of that "pretty world" in which "Gallantry took its last bow."

Virtually all the white characters rooted in the South's rural culture—with the exception of Scarlett's father, Gerald, an attractively romantic stage Irishman—are absurdities. Ashley, the object of Scarlett's lust, which drives the plot from start to finish, and Ashley's aristocratic cousin and wife, Melanie, are sentimental antiques, like Victorian mantel clocks. All Scarlett's beaux, and her husbands before Rhett Butler, are panting puppies, who bark briskly about going to war in defense of "honor," "the Southern code," and so forth, and who return from it either dead or deadly dull.

The Old South posturing, the gentlemanly *this* and ladylike *that*, to which all these characters are much given, gets nothing but scorn from the figures who dominate the script. For Rhett Butler, the handsome Charleston opportunist, and for indomitable Scarlett, the pious, fiery War rhetoric is ridiculous, and they say so. Rather than an occasion for lapsing into nostalgic obsessions for the way things were—which, we are shown in Gerald O'Hara, has the power to drive men insane—postwar hardship is a school of forgetting all that, a training for prosperity in the industrializing and urban postwar South. Indeed, Scarlett, despite her occasional wistful mutterings about Tara, and certainly Rhett, in all they do and utter, perfectly embody the ideals of New South greed, enterprise, and cunning—the ideas, that is, officially despised by the Southerners who kept waving the muddy banner of the Lost Cause long after battlefield defeat, and invented a myth of Knights and Ladies Fair to justify doing so, and whose spiritual heirs, with considerable irony, have loved *Gone With the Wind* since it came to the screen in 1939. (It was incidentally the last of the Old South films purporting to deliver "accurate" pictures of plantation life, a hugely popular and successful moviemaking trend launched in 1915 by D. W. Griffith's *Birth of a Nation*, and abruptly suspended by Hollywood at the outset of the Second World War. The United States' military censors, a film scholar has written, decided that "[f]ighting a war for the free world and democracy" and against racism precluded at-home "celebrations of the plantation.")

When I look back on our moviegoing episode over that long distance, it seems to have been a rather mean exercise in debunking, by two Southerners abroad and anxious about where we'd come from. For no matter how loud our grousing about the film, *it* was really not the problem. The *reception* of it was—the national appropriation of it by propagandists of all sorts, ranging from the Yankee students who laughed at the way we pronounced Anglo-Saxon, to the hucksters of moonlight-and-magnolias memorabilia, to those appalling Southerners who believed civil rights was leading straight to a mass ambush of white women of the South by blacks, just like the scene in *Gone With the Wind*, where Miss Scarlett is set upon and nearly raped outside Atlanta. It did not matter that the assailants in this scene are white. They *had* to be black; racist imagination demanded that they be black, simply because they (unlike director David O. Selznick) could not imagine a white man raping a white woman. Only black men were capable of rape. This genial, good ol' boy racism, that assimilates every fact and deforms every incident into its maliciously eroticized paradigm, is as sickening to remember and write about now as it was to witness in the Southern 1960s.

I claim no high moral ground. The issue of race has always been, in Southern imagination whether benighted or enlightened, including my own, an *aesthetic* consideration; quite literally a matter of

1 Introduction

In the now classic monograph by Pimentel and McClellan entitled *The Hydrogen Bond* (309), the brief section concerned with the C—H groups was entitled 'Do C—H Groups Form Hydrogen Bonds?' The conclusions, for the compounds and classes of compounds considered, were as follows:

trichloromethane. 'The evidence in favour of association of chloroform with bases is conclusive. The evidence that this association is of the hydrogen bonding type is substantial, and it is consistent with the statement that chloroform forms hydrogen bonds, at least with such strong bases as pyridine and triethylamine. Thus it must be concluded that, in this instance, a C—H group can serve as a hydrogen bonding acid.'

methanenitrile. '. . . the evidence for hydrogen bonding in HCN is strong if not conclusive.'

alkynes-1. 'Evidence for hydrogen bonding of acetylenic C—H bonds is good, but it is not voluminous.'

aldehydes. 'There remains uncertainty concerning hydrogen bonding by the aldehydic C—H bond.'

A considerable amount of study of intermolecular interactions involving C—H groups has been carried out since this monograph appeared in 1960, the majority of reported researches having dealt with the properties of trichloromethane and the alkynes-1. The conclusion to be drawn, after more than a decade of further study, is that Pimentel and McClellan were essentially right, although perhaps a little too pessimistic.

Interactions involving C—H groups have been a matter of considerable interest and occasional controversy. For example, it is agreed that molecules of methane, CH_4, interact with one another and with other molecules virtually exclusively by means of London forces, i.e. by means of induced dipole–induced dipole interactions. The case is similar for a symmetrically substituted methane molecule such as tetrachloromethane, CCl_4.

However, the asymmetrically substituted chlorinated methanes—H_3CCl, H_2CCl_2, $HCCl_3$—all bear permanent dipole moments and are expected to interact with polar molecules via dipole–dipole and dipole–induced dipole mechanisms as well as by London forces. The dipole moments: $6\cdot47 \times 10^{-30}$, $5\cdot34 \times 10^{-30}$ and $3\cdot50 \times 10^{-30}$ C m, respectively, would lead one to expect a decreasing energy of interaction. While energies of interaction involving these

color. But, then, perhaps no Southern social morality exists—only a regional, class-determined aestheticism masquerading as universal ethics. White Southerners from "old families" have often been portrayed as overrefined dandies and *flâneurs*—Tennessee Williams certainly thought that's what we were—and perhaps our critics are right. Snobbery toward Southerners of several varieties—*nouveaux riches* with their Tara-type mansions (built circa 1950) in Birmingham and Atlanta suburbs, no-count cousins, the plump matriarchs and decayed belles who had become bearers of the Old South mythology when the Great Depression destroyed the spirits of the men across the South—certainly informed our criticism of *Gone With the Wind*. But there was *something else* at work, at least in my mind—an anxiety that I did not, perhaps could not, acknowledge at the time. It was the thought—intolerable to me then, as I was trying to wrest myself out of Old South obsession—that the film's brief introductory invocation of the antebellum world of my ancestors, with its gallantry, civility, and devotion to the land, might be true to fact.

That would merely revive its danger to me, by making that oppressive version of the Old South inevitable, *inextricable* from my history. For if the *New* South and the *New Deal* South were clearly historical epochs, situated in time and comprehensible as eras of conflict and change—hence open to review and criticism in my mind— the *Old* South had become for me a bright zone *outside* the turmoil and procession of history's consecutive eras—a spot in primordial time, from which come our oldest myths and narratives of identity, and to which we return (in all the great epics of the West since *Gilgamesh*) to recover the selves we lost when we fell from that holier time into this profane one.

I rarely thought of this again until after Aunt Vandalia's death. I had hardly begun to plow through the heap of her genealogical papers when it occurred to me how *many* Souths had met and mingled in the life of this one Southern woman. There was the rural tobacco-planting culture of the oldest South, rooted in the British imperial outpost of Virginia—a culture old-fashioned by the eighteenth century (at least by the standards being set by British and northern American industrialization), but that provided the basic ideals of learning and life for the emerging South's elite. After 1830—in a

molecules are certainly larger than those involving CH_4 or CCl_4, they in fact increase in the order given, contrary to expectation. This lack of agreement with the predicted ordering of energies led to the postulate of a further type of interaction which was labelled as 'hydrogen bonding' since molecules such as H_2CCl_2 and $HCCl_3$ were found to exhibit some types of behaviour similar to well characterised hydrogen bonded systems involving molecules such as alcohols. and amines. A discussion of the nature of these 'extra forces' is given in Chapter 6.

An Historical Survey

It was in the mid-1930's that it came to be recognised that C–H bonds participate in a form of intermolecular interaction which is known as hydrogen bonding. At the time (1937) that Glasstone presented the results of his polarisation measurements which showed that compounds such as the trihalomethanes, HCX_3, form 1:1 complexes with donor molecules* such as ethers and ketones [4], doubt was expressed about the nature of the attractive interaction involved and as to whether or not the C–H groups were participating. A short time later, Gordy was able to show that the vibrational behaviour of the C–H group in trichloromethane [154], tribromomethane [154] and phenylethyne [363] is affected when these substances are dissolved in electron donor solvents. Thus, the fact that certain compounds can interact with donor molecules via a C–H bond was established, and the technique of vibrational spectroscopy has been exploited a great deal since then in the study of these complexes. A relevant survey was reported in 1963 by Allerhand and Schleyer [7] in which the effects on the C–H stretching vibration of a variety of molecules in the presence of dimethyl sulphoxide and azine were measured.

Many other techniques have been used for the detection and characterisation of C–H hydrogen bonds; these are mentioned and discussed briefly in Chapter 2 as well as in the appropriate Sections of Chapters 3 to 5. The principal technique after vibrational spectroscopy has been nuclear magnetic resonance spectroscopy, particularly in the last ten to fifteen years. Since the signal from a hydrogen nucleus is particularly sensitive to changes in the nucleus' electronic environment, hydrogen bond formation, even though weak, is detected easily. A recent NMR spectral study of $HCCl_3$ complexes which is representative of the present state of the art is that by Homer *et al.* [178]. The alkynes-1 have been of particular interest to Russian workers [see 301–304]; a more recent study is found in Reference [82].

---

* The use of the terms *donor* and *acceptor* in the literature of hydrogen bonds has been inconsistent and confusing. In this monograph, the more general sense of *electron* donor and acceptor has been used rather than *hydrogen* donor and acceptor. Thus, the C–H group is always a part of the *acceptor* molecule.

2 Experimental Techniques

In this chapter, a brief account is given of the various experimental techniques which have been used for the detection and characterisation of hydrogen bonds involving C–H groups. For the interested reader, a more detailed description of these and other techniques is given in reference [309]. Therein, Pimentel and McClellan suggest that these may be divided into two broad classes:

(a) those techniques which provide evidence of some sort of bond formation; the nature of the bond formed must be inferred from other information;
(b) those techniques which provide evidence that a new bond joining X–H and D specifically involves the hydrogen atom of the X–H group.

The former are usually general physical properties of the substances involved; these properties usually change in some way when a specific intermolecular interaction occurs. The class (a) techniques are described in Section 2.1.

The class (b) techniques are essentially two in number: vibrational spectroscopy (infrared and Raman), and nuclear magnetic resonance spectroscopy. These forms of spectroscopy are sensitive to changes in each part of a molecule, rather than in the molecule as a whole, and hence can provide evidence for the presence or absence of participation of a particular hydrogen atom in a bond formed between two molecules. They are described in Sections 2.2 and 2.3, respectively.

2.1 General Physical Properties

2.1.1 Vapour Pressures

When two liquids are mixed to form a solution, Raoult's Law states that the vapour pressure of each is directly proportional to its mole fraction; for such an ideal solution, then, the total vapour pressure varies linearly with the mole fraction of either component. However, in the case where there is an attractive interaction between molecules of the two substances which is significantly greater than the attractive forces between like molecules, then the vapour pressure of each component, and consequently the total vapour pressure, is less than that predicted by Raoult's Law. Such negative deviations from Raoult's Law are found in mixtures of trichloromethane with azine, as discussed in Section 3.1.7.5.

2.1.2 Azeotrope Formation

Arising from negative deviations of binary liquid mixtures from Raoult's Law

3

(Section 2.1.1), vapour pressure vs. composition graphs which exhibit minima are indicative of liquid mixtures which form maximum azeotropes. Hence, the observation of a maximum azeotrope for a binary mixture establishes the presence of attractive interaction between the components. However, when the boiling points of the components are greatly different, no azeotrope will be found, even though attractive forces between the components are present. For examples of azeotropes involving C–H hydrogen bonds, see Sections 3.1.7.4 and 3.4.5.4.

2.1.3 Second Virial Coefficient, *B*

The virial equation of state is a means of accounting empirically for deviations of real gases from ideality:

$$PV/nRT = 1 + B(n/V) + C(n/V)^2 + \cdots \tag{2.1}$$

The second virial coefficient, B, is the largest and is generally found to be negative since molecules of a real gas exert attractive forces on one another (unlike those of a 'perfect' gas). Such behaviour is described by the Berthelot equation (e.g. ref. 232); however, for gases or gas mixtures in which hydrogen bonding occurs, B will be more negative than predicted by the Berthelot equation. Typical results are presented in Sections 3.1.7.13, 3.5.4 and 4.3.1, confirming this expectation.

2.1.4 Solubility

The solubility of an electron acceptor compound is expected to be significantly greater in a solvent whose molecule contains an electron donor group than in one whose molecule does not. Thus, if a C–H group in a molecule is able to participate in hydrogen bonding, the solubility of the substance should be greater in electron donor solvents than in non-donor solvents. The situation is much different, however, with donor solvents which are also capable of participating in hydrogen bonding as acceptors, such as alcohols, acids, and some amides. Then, hydrogen bonding with the C–H group of the solute usually does not occur because solvent self-association predominates. Hence, the solubility may be *less* in such solvents than in non-donor solvents. Examples of these effects are given in Sections 3.1.7.14, 3.2.5, 3.4.1.1 and 5.2.5.

2.1.5 Freezing-point Diagrams

The diagram of freezing point as a function of mole fraction for mixtures of two substances normally has the form shown in diagram (a). However, when the substances combine to form a complex, a maximum in the diagram is observed

region beleaguered by rising antislavery sentiment in the North, the mechanization of the South, and the profound doubts of many Southern intellectuals and ministers and politicians about the rightness of slavery—these ideals were condensed into the Old South notions of chivalry and belles and such. Though born to New South wealth earned in banking and trade and land speculation, Vandalia was reared (like everyone else in my postbellum family) in an atmosphere saturated with romantic notions of self and society that had survived the ruin of the Old South ideology's most disastrous offspring, the Confederacy, and lived on in her mind as ghostly, compelling memories.

Trying to capture the cultural tone, the flavor, of Old South myth in our region before the Civil War is, at once, the lightest and most arduous of tasks. A sizable university library is not far from anyone living in urban North America. There, the seeker can find row upon row of sound, modern historical and archaeological and architectural and broad cultural studies about the life of the mind in our region before 1861. And that's before reaching the vast *contemporary* literature: sermons and stump speeches, writings poured by low-country literati into such influential journals as the *Southern Review* (1828–1832) and *Russell's* magazine (1857–1860); the fiction and essays of William Gilmore Simms, John Pendleton Kennedy, and, most important of all—though his stories and poems invoke the *atmospherics* of Southern chivalry's darker side, not our history or people—Edgar Allan Poe; the travelogues of the great landscape architect Frederick Law Olmsted (who gave us New York's Central Park) and many other observers, and a pile of humorous antebellum books with such titles as *Odd Leaves from the Life of a Louisiana Swamp Doctor.*

The beauties and delights to be found while climbing this tall escarpment of books by and about antebellum Southerners made the job interesting. But even while I enjoyed the flowers that grow here and there on that steep slope of tomes, the strictly *literary* character of the enterprise was drying up my will to continue. Surely not because I have anything against literature and historiography, but merely because my reading was taking me no closer to whatever truth about dwelling or durable goodness may have been created on Southern terrain in the decades just before the Civil War.

at the mole ratio corresponding to the composition of the complex: e.g. 2:1 in diagram (b). In favourable cases, this technique is quite sensitive and has provided evidence of association when other techniques do not. See Section 3.1.7.12.

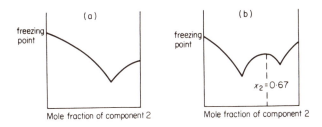

2.1.6 Enthalpies of Mixing

When hydrogen bonds are formed upon mixing two substances, the enthalpy change on mixing, $\Delta_m H$, is found to be more negative than it is for substances which do not form such bonds. Measurements of $\Delta_m H$ are reasonably straightforward to carry out, and a considerable number of values are now known, many of which indicate the formation of hydrogen bonded complexes. Further discussion of the results is to be found in Sections 3.1.3, 3.2.3, 3.3.1.1, 3.4.1.3, 3.5.5, 4.2 and 5.2.3.

2.1.7 Complex Dipole Moments

It is well known that the dipole moment of a molecule may be different in solution from its value in the gas phase and, indeed, may differ from one solvent to another. Several measurements have been made of mixtures of an electron acceptor and a donor in the expectation that the dipole moment of a hydrogen-bonded complex would prove to be equal to the vector sum of the moments of the two components. Unfortunately, the very process of formation of a hydrogen bond involves a redistribution of the electric charges within the component molecules, as well as between them (see Chapter 6), so the moment of the complex is generally not equal to the value predicted. For further discussion, see Sections 3.1.7.1, 3.2.8.1 and 4.3.1.

2.1.8 Viscosity

It has been found that measurements of the viscosities of binary liquid mixtures can be made to yield some information concerning the formation of complexes between the components [249, 286]. In particular, the excess viscosity, η_E, the deviation of the observed viscosity of a mixture from that expected from a linear

The thing to do, I decided, was again to stop reading and to find *one place* where my kinsmen had dwelt for generations before and after the Civil War, and which they had transformed and marked with their visions in the duration. In the course of sounding out the colonial experiences of my ancestors, I had found such evocative places in Virginia. In the middle Saluda River region of South Carolina, farmed by my early republican kinfolk, I had found spots similarly heavy with memory. But where was I to go to find legible traces of an *antebellum* existence constructed by my kin? I did not know.

It was during a telephone conversation, as I was grumbling about the dead end I seemed to be facing, that Joe Chandler asked: "Why not Mayesville?"

Like so many other tips my North Carolina cousin had given me along the way, this suggestion made instant sense. Much is known about Joe's ancestor Matthew Peterson Mayes, who, in 1820, founded the planting community, some eight miles outside Sumter, South Carolina, and where, in 1878, he died. Matthew was born wealthy and well-connected in 1794 near Emporia, south of Petersburg, Virginia, from the branch of Daniel Maies, brother to my seventeenth-century ancestor William. After the death of Matthew's father in 1796, he was reared by his mother—a kinswoman to Thomas Jefferson—at Petersburg, in the same Appomattox River vicinity where Daniel's sons had farmed and prospered for more than a hundred years before the Revolution.

Upon reaching maturity in the early republic, as the Northern seaboard states were quickly industrializing and becoming more urban, Matthew found himself in a world deliberately *holding back*. The tobacco planters of the tidewater and fall line, great and small, preferred to go on planting tobacco as they had throughout Virginia's decades as a royal colony; and, during Matthew's boyhood at Petersburg, they were still using their political and more informal, feudal powers to halt the southward and inland march of industrial enterprise. From the riverside overlooks near Petersburg, young Matthew could have beheld a scene that had changed very little in nearly two hundred years: a downstream drift of tobacco-bearing barges from the inland plantations, the crops unloading and warehousing at river stations, and reloading on sloops and schooners bound for Northern and overseas markets.

But by the time his home state had let in cotton, Matthew Peterson Mayes was gone. Everything known about his early years indicates that, from the age of reason onward, this intelligent young man had no intention of staying put in the deliberately *retardataire* economy of eastern Virginia. In 1814, at age nineteen, he left home to fight in America's second war of independence from Britain; and thereafter set himself up in successful retail trade in the dawning commercial and industrial town of Raleigh, North Carolina. But within only a few years, Matthew was gone again—this time to his rich new wife's district in South Carolina's luxuriously fertile low country. His bride died almost immediately, whereupon Matthew quickly married Henrietta Warner Shaw, daughter of yet another wealthy, well-entrenched local family. In 1821, he permanently established himself and his second wife, and a large house and plantation, on the site of what would eventually be known as Mayesville.

Matthew Peterson Mayes had hardly arrived before he began imprinting his name and ideas on the place, with the deliberate forcefulness, and the careful avoidance of ostentation, of a Roman aristocrat establishing his *villa rustica*. He was not yet thirty years old.

That much is history. What intrigued me more than this rapid ascent, or even his subsequent career in politics or his substantial worldly success, was the long shadow Matthew Peterson Mayes cast on his descendants and relations and neighbors, who honored him, during his life, with the honorific title *Squire*, and, after his death, in writing and family legend, and in their choice of Matthew's conservative rural model as their own. Matthew Mayes had also left marks on the ground—a plan of settlement laid out on Carolina dirt, and the architectural legacies of a dynasty of farmers and merchants grounded in his tradition and this place—tangible evidence, that is, to which the portraiture bequeathed to us by history and genealogy can be compared.

Exactly what was waiting for me in the town created by Matthew Mayes, I had no clear idea—but I was sure *something* would be there. And so it was, one cool afternoon in late autumn, that Joe Chandler and I wheeled out of Raleigh, where my cousin works for his governor, and headed south on Interstate 95, which today lies along roughly the same route young Matthew had taken, years before the Civil War, to his new home in South Carolina. After a swing by

dependence of viscosity on mole fraction, should be positive in mixtures where hydrogen bonds are formed, since the resulting complexes are larger and less mobile than the component molecules. The relevant results are found in Sections 3.1.7.15, 3.4.1.2, and 5.1.3.

2.1.9 Refractive Index

Pimentel and McClellan [309, p. 55f] mention that changes of refractive index have been observed in mixtures of hydrogen bonding substances, and that these could be significant provided that no extraneous hydrogen bonding species (such as solvent) are present. One such study, reported in Section 3.1.7.11, seems to bear out this expectation.

2.1.10 Electronic Spectroscopy

Although not a physical method for detecting hydrogen bonds, this technique is included here because it is, in general, not specific for a particular mode of complex formation, such as hydrogen bonding. As mentioned by Pimentel and McClellan [309, p. 158ff], a shift to higher frequency of the n $\to$ π^* transition of the donor molecule is often taken as indicative of hydrogen bonding, other spectral changes being more difficult to account for with confidence [see Sections 3.1.7.7, 3.5.2 and 4.2].

2.2 Vibrational Spectroscopy

Infrared and Raman spectroscopy are used in the determination of the characteristics of molecular vibrations. The most informative feature, the frequency of a vibration, is determined by the masses of the vibrating atoms, their positions in the molecule of interest, and the forces restricting the atoms to their respective positions.

In the simple approximation of a diatomic harmonic oscillator, the vibrational energy would be given by:

$$E_\mathrm{v} = \frac{h}{2\pi}(v + \tfrac{1}{2})(k/\mu)^{1/2} \tag{2.2}$$

where h is Planck's constant, v is the vibrational quantum number and may take any non-negative integral value, k is called the force constant and is a measure of the force tending to restore the atoms to their equilibrium positions, and μ is the reduced mass of the oscillator $(=m_1 m_2/(m_1 + m_2))$. Hence, the energies of transition between vibrational levels (as defined by the vibrational quantum number, v) are simply

$$(h/2\pi)(k/\mu)^{1/2},$$

Sumter to pick up Joe's first cousin Rutledge Dingle, assistant principal of the town's large high school, we crossed the few miles of
country to the town named after the kinsman of us all.

Had I expected Mayesville to be a snippet of Old South movie
footage, which I hadn't, our trip would have been a crashing disap-
pointment. The town—if that's the right word for it—is today a scatter of houses, ranging in style from modest to egregiously grand, a
shut-down gin and derelict hotel, a couple of churches, some tiny
juke joints rattling with hot dance music on a Saturday night, and a
few other buildings, broadcast over a wide area without edge or center or defining axis. The most conspicuous visible feature in
Mayesville is a huge absence: the bare strip of nothing, long and
wide enough to land an airbus on, that was opened in 1853 for the
railway, and was abandoned years ago. Nowadays, this vast interval
comes alive only occasionally; Joe, Rut, and I arrived just as one
such event was winding down. It was the annual parade and barbe-
cue hosted by the Mayesville town government, which had drawn,
councilor Eugene Davis told us, thousands of local folk—all of them,
like the council, its mayor, and the majority of the district's current
population, black.

Melting away back into the town, or retreating into Gladys's
Lounge and the other little saloons on the strip's edge, the festive
crowd left behind a deep emptiness, in which the silent architectural
witnesses to Mayesville's vanished white dominion seemed especially bereft. In the archipelago of widely spaced buildings put up
along the railway during the town's heyday, a century ago, the long
two-story Kineen Hotel stands empty now, its once-beautiful lobby
and wide ground-floor shops in ruin, floors littered with dusty glass
fragments from shattered windows. Across the expanse from the
Kineen, the handsomely roughened Roman arches of the J. F. Bland
Sale and Feed Stables—purveyor of mules for plowing the fields,
then hauling the yields of cotton, corn, tobacco, and soybeans to the
Mayesville depot—have been boarded up. The old stores that once
constituted the town's short commercial district, some of them with
their Victorian finials and fancy brick parapets still intact, are mostly
deserted now, or burned-out shells reclaimed by trees and bushes
growing up through fractured floors.

On a few streets lying near the railway wasteland stand the

the fundamental, and multiples thereof, (known as overtones). For most purposes, the fundamental vibrations are the most important.

For real molecules (i.e. anharmonic oscillators) and larger molecules, the situation is more complex, although qualitatively similar. In general, larger molecules have more vibrational modes, and each mode involves *all* the atoms of the molecule. Fortunately, however, many of the vibrational modes may be assigned *primarily* to the stretching of a particular bond or the deformation of a particular fragment of the molecule; thus it is possible to speak, say, of a 'C–H stretching mode frequency' even though this is not strictly correct.

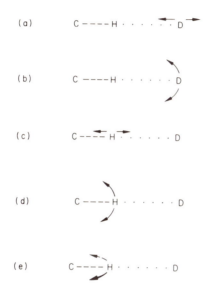

Fig. 2.1 Some vibrational modes of a C–H hydrogen bonded complex.

The vibrational spectrum of the acceptor molecule, and, to a smaller extent, that of the donor, is considerably modified by hydrogen bond formation. Since the vibrations involving the hydrogen atom of interest are identifiable in the acceptor spectrum, and distinct from those not involving that atom, infrared and Raman spectroscopy are class (b) techniques as outlined at the beginning of this chapter; that is, they provide evidence as to whether or not the hydrogen atom is involved in bond formation.

The vibrational modes which are of greatest interest in connexion with the hydrogen bond are shown in Figure 2.1. Mode (a) represents the stretching of the hydrogen bond itself. Although this mode has been observed for several strong hydrogen bonds, its low frequency (<250 cm^{-1}) has so far precluded its observation in the weaker C–H hydrogen bonds. Mode (b), representing bending of the hydrogen bond, probably has an even lower frequency and has also not been observed.

The other three vibrations illustrated are all vibrations of the acceptor molecule alone, but these are substantially perturbed by the influence of the donor molecule and hence indirectly indicative of its presence. Upon formation of a hydrogen bond, mode (c), the C–H stretch, is found to shift to lower frequency. In addition, the width of the peak, $w_{1/2}$, increases, and the intensity increases many-fold in the infrared spectrum but not in the Raman spectrum. The behaviour of mode (c) is the most frequently used characteristic in the determination of hydrogen bonds.

Modes (d) and (e), representing bending of the carbon–hydrogen bond relative to the rest of the molecule, will be degenerate if the acceptor molecule is symmetric about the C–H bond axis, as in $HCCl_3$, HC≡CH, etc. These modes are shifted to higher frequency upon hydrogen bond formation, with only slight changes of half-width and intensity.

The vibrational spectrum of the donor molecule is also affected by hydrogen bonding, although it is usually only the donor group which shows a significant effect.

Applications of vibrational spectroscopy are given in Sections 3.1.5, 3.2.6, 3.3.1, 3.3.1.2, 3.4.2, 3.5.2, 4.1.1, 4.1.2, 4.2, 4.3.2, 5.1.6, and 5.2.6.

2.3 Nuclear Magnetic Resonance Spectroscopy

Several aspects of nuclear magnetic resonance (NMR) spectroscopy have been found to be of value for the detection and characterisation of hydrogen bonds. Of these, by far the most important is the change in chemical shift of the nucleus of the hydrogen atom involved in a hydrogen bond. Of lesser importance are the effects of hydrogen bonding on spin–spin coupling constants and on relaxation times.

The NMR experiment involves measurement of energy from a radio frequency signal by a resonance of a sample contained within a very strong magnetic field. The resonance condition depends upon the relative values of the magnetic field strength, B_0, and the frequency of irradiation*. The measurement is of chemical importance because the magnetic field experienced by each nucleus is modified by the electron environment of that nucleus. The result, is that, for a given nucleus type, say ^1H, and for a given magnetic field strength, the irradiation frequency appropriate to cause resonance differs slightly from one electron environment to another, and hence from one chemical environment to another. The differences in frequency of resonance are called chemical shifts.

One of the most sensitive measures of a change in electronic configuration is the NMR chemical shift. Any perturbation of the electron(s) about the proton of the A–H group results in a change in its chemical shift. Hence, it is

---

* The reader is referred to Lynden-Bell and Harris' book *Nuclear Magnetic Resonance Spectroscopy* for a brief and lucid explanation of NMR spectroscopy.

remaining monuments of the closely knit white families that oper-
ated Mayesville—most conspicuously, as we might have imagined,
the mansions of the family compact founded by Matthew Peterson
Mayes. The boxy, becolumned house now standing among mossy
trees on the site of Matthew's original residence—it burned to the
ground in 1901—was realized in 1950 by James Edgar and Katie
Beaty Mayes according to the streamlined, rather suburban "Southern Colonial" canons of the day. Others were erected, predictably,
during the era of Mayesville's greatest prosperity, after the Civil War,
on the foundations of older structures demolished or severely tai-
lored to make way for new architectural celebrations of abundance.
Motoring among the old Mayes houses, we were not in the Old
South, but deeply in the Old South *mystique* of elusive elegance.

The most flamboyant of these dwellings—and the one I imagine
most non-Southerners would think the most quintessentially *Southern* of all—was built by R. F. DesChamps in 1910, later becoming the
property of William Rhodes Mayes. It is a specimen of what has been
fulsomely called the "Southern Classical Greek Revival" style—
though like many other similarly spectacular houses built elsewhere
in the United States during the Gilded Age, it reflects less a Southern
idea of design than the era's international taste for excessive decorative effect at the expense of architectural common sense. The result,
as we have it in the DesChamps (or Mayes) home, is a comfortably
ample house uncomfortably married to a huge curved semicircular
portico held aloft by gigantic columns in a wildly florid Corinthian
order. "Blossoming magnolia and tulip trees complete the imagery
for a classic Southern home," wrote a local critic in 1982, shortly
after the deteriorating structure was shored up by its current owner.
"The Mayes House, in its restoration, maintains the classic standard
of Southern architecture that disappeared shortly after the Depression." This is a curious description of a monstrous stylistic salad
freely pastiched from the Academic Classicism of the early twentieth
century and antiquarian fantasia.

Everything I'd learned so far from Mayesville about itself—the
mixed, peculiarly sad message of past health and present decline—
could be said about any one of a thousand small Southern towns.
Very many of them had been seduced by cotton and the railroads,
then abandoned by them, and finally annihilated by boll weevils in

not surprising that the formation of a hydrogen bond, however weak, is easily detectable from the NMR spectrum.

It is considered [389, p. 85f] that the principal contribution to the change in shift is a reduction in the shielding of the proton; this occurs under the action of the electric field of the donor species which tends to draw the hydrogen nucleus away from its electrons and also to polarise the electrons of the A—H bond so as to reduce the electron density in the vicinity of the proton. A secondary effect which is occasionally of importance is the magnetic anisotropy associated with the donor molecule. The principal effect always operates in such a way that it reduces the shielding of the proton and causes its resonance to be observed at lower magnetic field strength (higher frequency). The secondary effect usually results in an increase in shielding; in a few cases, this contribution is the larger and the observed resonance moves to higher field.

Two distinct kinds of chemical shift measurements have been made on complexes of acceptor molecules containing C—H groups. The majority involve measuring the difference in the ^{1}H shift of the C—H group in a donor solvent and in an 'inert' solvent. This has been done in the hope that the magnitude of such a shift difference, termed here a *solvent shift*, would give a measure of the strength of the interaction with the donor molecule. Some studies have measured the 'dilution shift', which is the shift difference when the 'inert' solvent is the acceptor itself.

The second kind of study has determined the difference of the ^{1}H shift between the *complexed species* (in a defined medium) and the acceptor molecule in some 'inert' solvent. These shift differences, termed here *complex shifts,* are indicated in the appropriate tables by italics.

The difference between the *solvent shifts* and *complex shifts* is significant in view of the fact that C—H acceptor molecules form rather weak complexes. The shift in a donor solvent results from acceptor molecules in more than one environment. In addition to 'complexed' molecules, there will be 'free' or 'uncomplexed' molecules. Since the exchange of acceptor molecules between environments is very rapid, the observed spectrum consists of a single signal whose position is a population-weighted average over the various environments.

In the simplest case of two environments, 'free' and 'bound', we have the following diagram showing the relationship between *solvent shifts* and *complex shifts.*

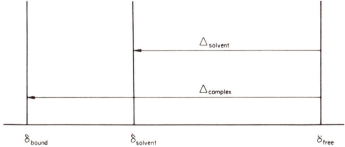

The *complex shift* is likely to be of greater interest than the *solvent shift* because the former is a measure of a property in a well-defined state, whereas the latter is not. The *solvent shift* is in a position analogous to that of the enthalpy of mixing (see Section 3.1.2) in that it contains a combination of a property of the complex with its formation constant.

If it is possible to make the assumption that, in a donor solvent, acceptor molecules are present in only two environments—viz. 'free' and 'bound', then $\Delta_{complex}$ and $\Delta_{solvent}$ may be related to each other through the complex formation constant.

$$A \quad + \quad D \quad \xrightleftharpoons{K_\phi} \quad AD$$
$$\text{acceptor} \quad \text{donor} \quad \text{complex} \tag{2.3}$$

$$K_\phi = \frac{c_{AD}}{c_A c_D} \tag{2.4}$$

where c_i is the concentration of species i. The *complex shift*, $\Delta_c (= \delta_{AD} - \delta_A)$, is related to the solvent shift, Δ_s, by the ratio

$$\frac{\Delta_c}{\Delta_s} = \frac{c_{AD} + c_A}{c_{AD}} = 1 + \frac{1}{K_\phi c_D} \tag{2.5}$$

When the concentrations are expressed as mole fractions, then at high dilution, $c_D \rightarrow 1$. The relation above simplifies to

$$\frac{\Delta_c}{\Delta_s} = 1 + \frac{1}{K_\phi} \tag{2.6}$$

the 1920s and the Great Depression. The intense pathos, emanating from the derelict hotel and shops along Mayesville's railway corridor, is the usual melancholy of little towns. But melancholy was surely not the only tale Mayesville had to tell. I had still not found the antebellum South that our cousin Matthew had begun constructing on this ground almost 180 years before I got there, and which I knew must lie somewhere.

I might never have found it, had not Joe, Rut, and I decided we wanted some fried chicken, and stopped by Mayesville's Exxon station to eat. Or, to be perhaps too precise, had I not needed to go to the men's room. Upon emerging from it, I saw a framed map hanging on the wall in front of me. For an instant, I thought it was the same one kindly sent to me, in advance of my trip, by the Sumter County Genealogical Society—a hand-drawn scheme with dots denoting notable houses and churches and so forth, and captioned: "Mayesville S.C. 1984 Riding Tour Chart—may be used to view Mayesville Community from your car." Useful enough when one is on a "stately homes" drive-by, it said nothing to me about the Squire's plan for the town, if he had one, and it seems he did. Joe and Rut had told me a story passed down through generations of their family, to the effect that, in developing his plantation, Squire Mayes had eventually included three dead-straight *allées* radiating from his manor house to the residences of sons—a coherent, basic pattern easy to fill in as the population became denser, as it rapidly did. Like his cousin, General Samuel Mays, Matthew was from his earliest years a careful designer of his destiny. So why not his town?

To my surprise, the carefully pencil-drawn illustration in the Exxon station revealed exactly what the plan had been. Its author, a bright lad of the town named W. P. Nesbitt, had, in 1916, the sense of place the anonymous draftsman of my 1984 map lacked—an idea of the underlying order of Mayesville, now (and perhaps then) all but lost to view. The outer limit of the town is indicated by a circle—a familiar alternative to the square on South Carolina maps. At the top of this circle, just inside the perimeter, is the homeplace of Matthew Peterson Mayes. From this point, exactly as predicted by the family legend known to Joe and Rut, extend the radial avenues (streets by 1916) across the circle. One leads to the house of Dr. Junius Alceus Mayes, the Squire's first son, born in 1822 (and named after, respectively, a famous Roman orator and a notorious Greek political satirist). Another goes to the 1843 house of Thomas Alexander Mayes. (If there was a third *allée*, as local history and Mayes family memory assert, young Will Nesbitt knew nothing of it.) More interesting still, the lines drawn from Squire Mayes's home to his sons' houses, each one by the perimeter opposite, are roughly the same distance back to the patriarch's residence. On the center of the circle the cartographer writes: "colored buildings too small & numerous to set down." The area was almost certainly populated by blacks long before 1916—which, in turn, suggests that the Squire had determined the location of his family's slave quarter by the time he built Thomas Alexander's house: in the middle zone of the circle anchored by his own and his sons' dwelling places, hence conveniently equidistant from their plantations beyond the perimeter. It took only a few minutes of driving and walking around Mayesville to find evidence of the geometry glimpsed by Nesbitt under the many transformations of the site since 1820—and to see our mapmaker was right.

While thinking back later on the map at the Exxon station, I remembered a good story retailed by the Roman architectural theorist Vitruvius, in his hugely influential treatise *De architectura*, about the Socratic philosopher Aristippus—who, shipwrecked on Rhodes, found geometrical diagrams on the sandy shore, and exclaimed: "Let us be of good cheer, for I see the traces of man!" For wherever a geometrical figure is inscribed on landscape, humanity must certainly stand close at hand, and the strategizing, abstract intelligence unique to our kind. The figure need not be *perfect*; the scheme Matthew had in mind ended up skewed by the lay of Mayesville between the curiously named Scape Ore (or Scape Whore) Swamp and the Black River. Nor need it be *intact*. The cross-cut of the Seaboard Coast Line through Mayesville severed the avenue between the homes of the Squire and his son Thomas Alexander, Junius Alceus's house is gone, the few Mayeses who still live in Mayesville live where they like, and the notion of some foundational geometry to the town, if alive in folk memory, is now all but invisible. But, following Nesbitt's indispensable map, we can still make out under the lives lived and houses built by generations of Mayesville people the unmistakable, precise signature of Matthew Peterson Mayes, exactly where I'd hoped to find it: inscribed on Southern ground.

*   *   *

In 1930, as the South was sliding into the Depression and Mayesville on the way to its present dilapidation, Sallie Ruberry Burgess sat down at her typewriter in Greenville, South Carolina—much as Samuel Edward Mays was doing in Florida, in the same moment of economic uncertainty and anxiety—and began to compose a manuscript with an assuring, wrap-up title: "The Mayes Family of Virginia and South Carolina."

I suspect she intended to write a definitive book about the descendants of Reverend William of the Virginia Company. At first, genealogy always seems to be an easy and congenial study—a recollection of strong ancestors who recall us to duty and loyalty in the midst of calamity. At least it *seems* easy, until all the interesting family tales and notable names, hitherto stowed away in memory, have been written down, the dates and names in family Bibles have been neatly sorted—and, suddenly, contradictions and ambiguities regarding dates and places and kinship begin to emerge. The famous ancestor invoked in family stories all your life evaporates—it was all wishful thinking! Or the conniving wretch or fallen woman or family villain turns out to have been deeply wronged by your near kin, and it is years too late to mend the harm done. Or one discovers the daunting fact that he is—as Joe Chandler is—*his own cousin three times over.*

It's when passing over these first confusing, jarring rapids in the endless stream of genealogical work that aspiring family historians give up. Or, alternatively, keep going in one of two directions. One is upstream, into the rigorous analysis and exacting research that serious genealogy always entails eventually. The other is downstream, drifting along, collecting the butterflies of pleasant story, but without getting too scientific about what's caught in the net. Among my genealogically-minded kinfolks, Joe is definitely an upstream paddler. Sallie, on the other hand, was a drifter—not a bad thing to be, especially from the point of view of us nongenealogists. For, like Sallie Burgess, I appreciate a good story, and she gathered many a good one. In fact, without her "Mayes Family"—in which all the Mays and Mayes patriarchs save her own are disposed of in a little over two typed pages—we might have no extensive personal witness to the character and times of Matthew Peterson Mayes, who clearly awed her. And I would have only a couple of maps, the geometry he

3 sp³ Carbon

3.1 Trichloromethane, HCCl₃

The hydrogen bonding properties of the C–H group of the trichloromethane molecule have been studied more extensively than those of any other molecule. In view of the established opinion [7] that only compounds containing 'activated' C–H groups are capable of hydrogen bond formation, this compound should be one of the most favourable for such studies. Indeed, the polarity of the C–Cl bond is greater than that of the other C–halogen bonds, with the exception of C–F, and so hydrogen bond formation is expected to be favoured. Trifluoromethane, which should be much better still on this basis, has been very little studied, presumably because of the difficulty of handling the gas (b.p. 191 K).

3.1.1 Structure and Stoichiometry of the Complexes

In the main, complexes involving trichloromethane have been assumed to take on a 1:1 stoichiometry. Often, experimental conditions are chosen so as to favour the formation of such complexes only, lest the formation of higher complexes interfere with the interpretation of results.

However, evidence is available that higher complexes may also form. Table 3.1 (see p. 57) consists of a listing of the known higher complexes of trichloromethane. It is immediately evident that these complexes break down into broad categories.

The first of these categories comprises complexes with electron donor molecules containing more than one donor site. In these cases, it is possible for each donor site to interact with a trichloromethane molecule, presumably independently of the other sites in the same donor molecule. Indeed, the formation of a complex at one donor site should have little or no effect on the process of complex formation at a second site.

An exception to the general case occurs with the donor 2,5,8,11,14-pentaoxapentadecane, $CH_3O(CH_2CH_2O)_4CH_3$. The enthalpy of mixing of this compound with trichloromethane shows a maximum corresponding to a 3:1 species. The same study [406] showed that 2,5-dioxahexane, $CH_3OCH_2CH_2OCH_3$, forms only a 1:1 complex. These exceptions suggest that only alternate oxygen atoms of such polyethers are available for forming complexes with molecules of trichloromethane.

A similar situation has been found in dielectric loss measurements of trichloromethane with the silatranes [74] (numbers 11 and 12 in Table 3.1). These donor molecules, containing four or five potential electron donor sites, are thought to form complexes no higher than 2:1. In these cases, it may be that

11

steric and electronic repulsion between $HCCl_3$ molecules prevents more than two
of them from interacting simultaneously with each donor molecule.

The second category of higher complexes involves interactions of trichloromethane with molecules containing only one electron donor atom. In most cases
such complexes involve two trichloromethane molecules and a single oxygen
atom. It has been suggested [254, 18, 83, 396] that the $HCCl_3$ molecules may
interact with each of the nonbonding pairs of electrons associated with an oxygen
atom. While this type of complex would suffer from fairly severe mutual repulsion
between the trichloromethane molecules, the evidence from infrared spectroscopy
shows that the carbonyl stretching vibrations of ketone donors may be found at
three distinct frequencies, corresponding to the species

$$C=O \qquad C=O\cdots HCCl_3 \qquad C=O\raise1ex{\cdots HCCl_3}_{\cdots HCCl_3}$$

$$\text{I} \qquad\qquad \text{II} \qquad\qquad\qquad \text{III}$$

Alternatively, the suggestion has been made that the 2:1 complex species takes
the form

$$C=O\cdots HCCl_3\cdots HCCl_3$$

$$\text{IV}$$

The evidence for form IV is not very strong. The energy of formation of IV
from II should be similar to that for dimerisation of trichloromethane itself. It
is seen in a later section that trichloromethane dimerises only to a small extent.
Hence, structure III seems more likely than structure IV.

The 2:1 and 3:1 complexes formed by trichloromethane with the tetra-
alkylammonium halide salts (numbers 28–32) are likely to be of the same general
type. That is, the halide ions have several nonbonding electron pairs available for
complex formation, rendering higher complex formation possible.

On the other hand, 2:1 complexes of $HCCl_3$ with amines (numbers 33 and
34) cannot involve interaction with more than one pair of nonbonding electrons
since the nitrogen atom has only one such pair. Here, it seems necessary to
invoke a structure of type IV in order to account for the bonding in these
complexes.

A study of the freezing points of trichloromethane/diethyl ether mixtures [40]
indicates the existence of several higher complexes (numbers 23–5). The reported
complex $HCCl_3 . 3(C_2H_5)_2O$ is particularly interesting since it seems to imply
interaction of trichloromethane by way of the quite polarisable Cl atoms rather
than through the H–C entity. Although evidence for such interactions has been
found for HCl_3 [327, 43, 41, 40, 42, 170], it has usually been felt that $HCCl_3$
interacts preferentially through the C–H bond (but see [145] and [327]).
Unfortunately, the freezing-point diagram can give no information concerning
the structure of the complex.

Wyatt [403] states that the freezing-point diagram for trichloromethane and propanone shows that only a 1:1 compound is formed. This finding is in contradiction to other results (Table 3.1, p. 57, numbers 14–7). This may be one of the many examples of the existence of different complex species in different phases, such as solid and solution phases.

3.1.2 Complex Formation Constants

3.1.2.1 1:1 Complexes

It is important when considering the process of association of molecules to know the extent to which the molecules exist in the form of complexes. This tendency to form complexes is expressed quantitatively as the *formation constant* for the equilibrium between the complex and its components. Hence, for the equilibrium expressed as

$$A + B \rightleftharpoons AB \qquad (3.1)$$

the formation constant, K_ϕ, is given by

$$K_\phi = \frac{\{AB\}}{\{A\}\{B\}} \qquad (3.2)$$

where $\{X\}$ is the activity of species X. It is most often the case that the activities in this expression are replaced by the corresponding concentrations; the necessary assumption is made that the ratio of activity coefficients of the uncharged molecules is near unity. Homer and co-workers provide a lucid commentary on the effects of this and other assumptions normally used in the derivation of formation constants from NMR spectral data [178].

The formation constants of 1:1 complexes involving trichloromethane are presented in Table 3.2 (see pp. 58–66). The substance with which the complex is formed is given as the donor. If a third substance is present, to serve as solvent, it is listed under the heading Solvent; a simple binary mixture is indicated by a dash in the latter column.

Several obstacles to the successful comparison and correlation of these constants are immediately evident. First and foremost of these is the fact that about half the data are reported with concentration units and about half with mole fraction (mol mol^{-1}) units. One might expect that concentration units would naturally result from studies in which a solvent is used, whereas mole fraction units would arise from studies of binary mixtures. However, no such practice seems to have been followed; instead, the choice of units appears arbitrary.

Second, although the magnitudes of formation constants are well known to be dependent upon temperature, there has been no attempt to report values of K_ϕ at a standard temperature, say 298·15 K. Instead, this parameter is treated as one at the discretion of the individual.

imposed on the site Mayesville, and some unconfirmed hunches about the sensibility of this man, and about the shape of the shadow he constructed and cast on the South Carolina land.

As presented in her unpublished memoir of 1930, Sallie Burgess's grandfather "kept open house and 'all at Matthew's loved to dwell.' He sheltered all in need, and many were given in marriage from his fatherly home." Under the dire circumstances prevailing in occupied postbellum Mayesville, he set an especially memorable example of stalwartness. "One letter in 1866 to a little grand-niece says 'We are learning the rudiments of this new life,' and how brave they were!" Both before and after the War, "[h]is wisdom and high principles made him a safe guide in all matters of Church and State. For many years he was an honored elder in the Old Salem (B.[lack] R.[iver]) Church, where as citizen and Christian, he performed his duty well. Modest and unpretentious (always reminding me of 'The Father of His Country'), he was 'one of Nature's noblemen—a true gentleman' . . . And how I loved to walk with him when he blew his tiny whistle, and all the chickens, ducks, turkeys and guineas would come flying," recalls his granddaughter of her visits to his estate in Mayesville, "for they knew his call.

"And what child could forget the wonderful drives with that stately couple Matthew and [Sallie's grandmother] to Brick Church with 'Case' up on the seat in front of the fine old carriage, and how he would drive up to a large oak, then climb down and open the door and let down the brass carpeted steps. Automobiles cannot touch the grandeur of those days! 'The Squire' wore a silk beaver in those days of old, but the 'old school' passed with him."

Across Sallie's rather relentless idyll of the Squire's reign, death sometimes stalks—often, interestingly, along the railroad tracks that maimed Matthew's landscape geometry. The death of Matthew's second wife, the mother of his ten children, came "during the fearful scourge of fever following the opening up" of the railway; so did the death of his son Samuel, who "died of fever when [the] railroad was built." The mechanization of Mayesville, by trains and, later, automobiles, *did* "touch the grandeur of those days," and its caress was baleful. But the most intimate blow to Mayesville's symbolic topography, for Sallie, came when the Squire's beloved mansion was destroyed by fire. "The great old brick smoke-house may be there

The third factor affecting the magnitude of K_ϕ, although not always recognised is the solvent, or medium. It has been pointed out [162, 276] that the formation constants which are desired for purposes of comparison and calculation are those for the equilibrium

$$A(g) + B(g) \rightleftharpoons AB(g) \qquad (3.3)$$

However, for the sake of experimental convenience, measurements are usually made of the equilibrium process in solution, where each reacting species is solvated to an, as yet, unknown extent:

$$A(solv) + B(solv) \rightleftharpoons AB(solv) \qquad (3.4)$$

It has been pointed out correctly that *this* equilibrium involves at least the rearrangement, and more likely the displacement, of part of the solvation sphere of each species involved in formation of the complex. Since the energetics of the change in solvation will be different for different solvents, we can expect the extent of complex formation, and hence the formation constant, to be different also. It should be mentioned that measurements made on binary mixtures, so far from eliminating the uncertainty attending the role of a solvent, may be even more difficult to interpret where the nature of the solvating medium is so dependent on concentration.

A fact which complicates most attempts to determine formation constants of complexes involving trichloromethane is the well-documented self-association of the substance. The formation constant for its dimer has been determined by several workers and is given in Table 3.2. Obviously, this competing equilibrium must be taken into account in any serious attempt to determine other formation constants. Fortunately, the extent of self-association is rather small, and so has little effect on measurements of other formation constants.

The constants for formation of complexes with two oxygen-donor bases— propanone and diethyl ether—have been the most extensively studied. As may be seen in Table 3.2, these studies embrace a considerable range of temperatures and solvents, and the results are reported in different units. As expected, the magnitude of K_ϕ decreases with increasing temperature, indicating a negative enthalpy of complex formation (see Section 3.1.4).

The results of Buchowski *et al.* [54] show that K_ϕ for the trichloromethane/ propanone complex is greater in hydrocarbon solvents than in tetrachloromethane or carbon disulphide by a factor of 2 to 3. Similarly, K_ϕ in the solvent CCl_4 is larger than that in a binary mixture of the components by a factor of 2. Evidently, the magnitude of a formation constant is strongly dependent on the nature of the solvent, and hence on the nature of the solvation of the interacting species. This effect was demonstrated quite dramatically by Olsen [291] who found the $HCCl_3$/dimethyl sulphoxide complex formation constant to have the value $13\cdot4\,dm^3\,mol^{-1}$ in cyclohexane solvent and $0\cdot07\,dm^3\,mol^{-1}$ in dichloromethane.

still and a few crepe myrtles, but most of the cedar grove is gone, while the beautiful grey moss drapes the old trees standing like silent sentinels on guard . . . " All this, Sallie concludes, she has written "for those who do not remember, some I copied from dear Cousin Sue's letters, and wonder like her, 'if those who read will touch it gently.'"

The critically attuned mind recoils from recollection of this sort, handling it roughly, if at all. To be sure, family memoirs are not the places to go if what's wanted is the raw stuff of historiography; and, for that reason, vast heaps of them languish unread, unused, in attics across the South, until thrown out by heirs who do not recognize these treasures for what they are.

For Sallie, like so many other writers of family narrative, thought mythically, not historically—enfolding every fact into a radiant human presence of the Squire, timeless as the Knight of the Wasteland in the fairy tales Essie De read to me at bedtime in Spring Ridge—the hero who frees the waters and restores the dry earth, inaugurating the return and establishment of civilized life to the ruined site. If such storytelling is famously unreliable in matters of fact, it is a source for the understanding of a certain *mentalité,* and deserves the regard we give any poetry similarly seeking to enframe origin, destiny, more profound levels of existence than those lying at the surface. Though certainly no Homer or Virgil or Malory, Sallie Ruberry Burgess was similarly driven by the urgency, as old as the West itself, to fashion a founding myth that would nourish an emotionally impoverished present, and to proclaim with the great celebrants of the past, as she does toward the end of her unpublished memorial: "What a heritage is left to us!"

But what, exactly, is the heritage that the Mayes family compiled and bequeathed in Mayesville? As I had already discovered, it wasn't an architectural or literary one, nor just a geometrical sketch on the ground, as interesting as that is. Had I come to this little town, looked, and left, I almost certainly would never have comprehended Mayesville's most important message. Things turned out rather differently, however, because Joe and Rut had decided to take me to visit their cousins Billy Dabbs and his wife, Lynda Mayes Dabbs, who were also cousins of each other—as are many men and women with roots in Mayesville.

Retired after a lifetime of farming cotton, the Dabbses welcomed us at the front door of the spacious house Billy had constructed without plans. We introduced ourselves—Mays, Chandler, Dingle—and Lynda said: "I guess I'm related to you all!" The genealogical conversation following our exchange of greetings, as we settled by the fire Billy had kindled against the chill in the air of that brilliant afternoon, would probably read like a strange, disjunct poem were I to try setting it down here in full; nor, I suspect, would this charming discourse make quite good sense to anyone born and reared outside a certain kind of Southern family. It's something Southerners just *do*, as a way of telegraphing recognition. Even as a small child, I knew exactly what Sister Erin, my grandmother, meant when she said, as she always did when the topic of her distinguished Alabama grandfather came up, that he was "the law partner of the vice-president of the Confederacy." Whether he was a good lawyer or a bad one, a man with personal eccentricities or none, were matters of no consequence to Sister Erin. His importance lay in his place within a certain historical tableau, less *person* than *site* on the mental map of who we are.

The opening conversation at Billy and Lynda Dabbs's house, which my grandmother would have understood, had to do with our *situation*. Who was whose mother, uncle, cousin? And how did Joe (whom the Dabbses had not met) connect to cousin William or Miss Emma Mayes or someone else—or did they connect at all, and, if not, why not? Well, there *was* this great-grandchild of a certain fifth cousin twice removed who married—who was it?—and nothing more was heard of her. Or him. But, come to think of it, *that's* probably the connection—via Squire Mayes's great-granddaughter Mary Frances Burgess, who married the Reverend William Bratton Steele Chandler, a Presbyterian minister. And, sure enough, it was!

I should add that never, then or later, did it occur to anybody to ask what anyone else in the room *did*. *What one does*—I was recalling gradually, having had no experience of such genealogical determination for some years—is quite unimportant; to ask the question would be impolite, even offensive. *Who one is*, however, is very important—or, to be more precise, *where one is* on the map of time and blood that families like those of Joe Chandler and Rutledge Dingle carry around in their heads. After the relations among Joe and Lynda, Rut and

Billy had been decided, and the exact link between me and the Squire was adjudged undecidable, or just too hard to figure out that day, we fell into a long chat about Mayesville people and places.

If I found Billy Dabbs's wonderful tales a little hard to focus on, it was because I was still intoxicated by what had gone before. One reason was the simple, sprightly beauty of such genealogical talk, its gamelike freedom—which, like other games, yields up pleasure only when everyone is following the same strict rules, and observing good manners. *Who one is*, in this context, has nothing to do with prestige or wealth. Those are things anyone can acquire, and are therefore interesting only to snobs and gossips. It has everything to do, however, with matters utterly fixed in time: descent, parentage, and place of birth. I was enchanted by the introductory interchange by Lynda and Billy's fire, and intrigued by the awareness that I was very close to the key that might unlock the door into the Old South I was looking for in Mayesville, but that still remained bolted against me.

And in fact, some days later and back at home, I turned again to the documents, historical and genealogical, about the Squire and his town; and heard them unlock. And the first step over the threshold took me first, not to Mayesville, but into the broad South Carolina fields of the early nineteenth century.

When Matthew arrived in the area with his first wife, the prominent rice and indigo planters among whom he now found himself—cotton was yet to come—were loosely focused in a district called Salem, a few miles away from the future site of Mayesville. If these Scots-Irish, solidly Presbyterian immigrants of the middle eighteenth century, had ever been as cantankerous and malicious as those encountered by Charles Woodmason, by 1821 they were as staunchly settled, respectable, and tight-knit as the Anglicans farther down-country. Matthew, who had been raised Baptist, threw in his lot with the Calvinists, joining the congregation of the Salem Black River Church, which had been founded in 1759. As he did in all else he attempted, and as Sallie Burgess continually reminds us, he immediately rose to prominence in the assembly; today his remains rest in its churchyard, under a plain headstone.

The Salem Church, like rural meeting houses across the antebellum South, followed the English parish-church precedent in providing a focus for a dispersed community's concerns—cultural and political,

Among the oxygen-donor bases, it seems to be generally true that an increase in electron-donor ability, such as in a series of phosphine oxides, results in a higher formation constant. In the case of nitrogen-donor bases, by contrast, the situation is not straightforward. For primary amines, there is no apparent relation between chain length (presumed electron donor ability) and K_ϕ. On going from primary to secondary to tertiary amines, there is a *decrease* in K_ϕ, again contrary to expectation. The moderately strong base azine exhibits a small complex formation constant, whereas the weak donor ethanenitrile is found to have a much larger value of K_ϕ.

The aromatic amines (anilines) should not be compared with the other nitrogen-containing compounds since there is good evidence [283] that their π-electrons act as donor in preference to the nitrogen electrons. Hence, the anilines should be considered as aromatic donors.

Halide ion donors give rise to formation constants which are substantially larger than those of complexes involving covalently bound halogen atoms as donors. Although in tetrachloromethane and ethanenitrile solvents, the tetra-alkylammonium halide salts exist predominantly in the form of aggregates (ion pairs, triplets, quadruplets, etc.) [117], the complexes may be regarded as forming between a trichloromethane molecule and a halide ion, the neighbouring counter-ion or ions having only a secondary effect. Once again, a more basic solvent (ethanenitrile) results in a smaller observed formation constant.

A curiosity which has been reported is a complex involving carbon as the *donor* atom to give C–H···C. The complex formation constant for trichloromethane, cyclohexyl isocyanide is virtually the same as that involving simply cyclohexyl cyanide, implying a similar electron donor ability of the $-C\equiv N$ and $-N\equiv C$ groups.

The fact that a linear relationship is found between pK_ϕ for $HCCl_3/RC_6H_4NH_2$ complexes and pK_a for the bases [397] is likely fortuitous, as is shown by the deviation of points for RC_6H_4NHR' complexes with trichloromethane. The acidity constant, measured in water, is in effect an indication of how readily the nitrogen atom may be protonated, whereas the $HCCl_3$ complex is known to form with the aromatic ring in anilines [283].

Bellon and Luis-Abboud have made a test of the effect of deuterium sub-stitution on the complex formation constants of trichloromethane [33]. To a system consisting of an inert solvent (e.g. heptane), an acceptor (e.g. 2-naphthol) and a donor (e.g. 1,4-dioxane), is added a definite amount of $HCCl_3$ or $DCCl_3$. The near ultraviolet absorption of the acceptor or the complex is much affected because of a shift in the complex formation equilibrium; the latter occurs because of the competing formation of trichloromethane/donor complexes. The authors claim that this method is very sensitive, and applied it, in all, to 34 systems of varying solvent, acceptor and donor. It was wished to measure the ratios K_D/K_H for the reactions

$$A + HCCl_3 \xrightleftharpoons{K_H} A \cdots HCCl_3 \qquad (3.5)$$

and

$$A + DCCl_3 \underset{}{\overset{K_D}{\rightleftharpoons}} A \cdots DCCl_3 \tag{3.6}$$

for a series of acceptors. The value of $K_D/K_H = 1 \cdot 01 \pm 0 \cdot 04$ seems to be independent of the nature of solvent, acceptor and donor. Thus, the mass of the hydrogen atom in trichloromethane seems to have no effect on complex formation. Indeed, it was found in addition that

$$\Delta H_D - \Delta H_H = (1 \cdot 3 \pm 1 \cdot 3) \text{ kJ mol}^{-1} \tag{3.7}$$

so that the enthalpies of formation do not differ significantly.

Wiley and Miller [400] have made the relevant observation that the reliability of equilibrium constants determined from spectral data is highly dependent on the fraction of acceptor molecules which is 'complexed'. This so-called saturation factor, s $(=[AD]/([A] + [AD]))$, may take values from 0 to 1. Values of K_ϕ are most reliable when based on spectral data covering as much as possible of the range $0 \cdot 2 < s < 0 \cdot 8$, and become rapidly and increasingly uncertain towards 0 and 1. They point out that this criterion has not been well understood in the past, and hence many results are of inadvertently low reliability since they cover unsatisfactory ranges of s. This is, indeed, sometimes unavoidable when complexes are so weak or so strong that the middle range of s is inaccessible.

3.1.2.2 Higher complexes

Several values of formation constants have been measured for complexes involving two molecules of trichloromethane and a donor molecule. In the case of 1,4-dioxane, possessing two donor sites, one would expect approximately $K(1:2) = [K(1:1)]^2$, since the donor ability of an oxygen atom will not be much affected by complex formation at the other oxygen atom. In fact, $K(1:2)$ proves to be about the same as $[K(1:1)]^2$:

$[K(1:1)]^2/\text{mol}^2 \text{ mol}^{-2}$	$K(1:2)/\text{mol}^2 \text{ mol}^{-2}$	Ref.
1·96	1·74	[18]
1·23	1·38	[251]

For the donor molecules containing only one oxygen atom, $K(1:2)$ is expected to be less than $[K(1:1)]^2$ if the donor ability of the oxygen atom is reduced by formation of the 1:1 complex, *or*, if the oxygen is rendered less accessible by the presence of the first $HCCl_3$ molecule. Alternatively, the formation constant of a complex of type IV (Section 3.1.1) will also differ from $[K(1:1)]^2$. Such a constant would likely be smaller than $[K(1:1)]^2$ since the dimer formation constant for trichloromethane is quite small.

$K(1:2)$ for the complexes with ketone are of the order of $[K(1:1)]^2$, demonstrating that the donor ability of an oxygen atom is not necessarily reduced by the existence of a 1:1 complex. However, dimethyl sulphoxide and oxolane exhibit much smaller values of $K(1:2)$. Presumably, one or more of the factors mentioned in the preceding paragraph are operative in these cases.

3.1.3 Enthalpies of Mixing

A number of measurements have been made of the quantity of heat evolved on mixing trichloromethane with another substance. Since the heat change is normally measured under conditions of constant pressure, it may be regarded as the enthalpy change for the mixing process, $\Delta_m H$.

This enthalpy change of mixing will be related to the enthalpy change of complex formation (see Section 3.1.4) through the concentrations of reactants taking part in complex formation. McGlashan and Rastogi [265] have shown that this relationship takes the form

$$\Delta_m H = \Delta_\phi H \left\{ K_x \frac{(1 - x_D)\xi_D}{1 + K_x \xi_D} \right\} \tag{3.8}$$

where x_D is the *stoichiometric* mole fraction of the donor, ξ_D is its *actual* mole fraction in the mixture, and $K_x = \xi_{AD}/\xi_A \xi_D$, the subscripts A and AD referring to the acceptor (trichloromethane) and the complex respectively. The form of this relationship must be extended if higher complexes are present in the mixture as well as the 1:1 species [265].

The values of enthalpy changes of mixing are not of much interest in themselves, since they combine two factors: the extent of complex formation, and the enthalpy of complex formation. Only when these two factors can be separated, such as through the equilibrium (formation) constant, can quantities of interest be determined or predicted. A cogent example would be the enthalpy changes resulting from the mixing of trichloromethane and of tetrachloromethane with a common donor such as diethyl ether. Such a study [25] has found $\Delta_m H \approx 2600$ J mol^{-1} in the former case, and $\Delta_m H = 500$ J mol^{-1} in the latter. Unfortunately, these results taken in isolation give no information about the relative importance of K_ϕ and $\Delta_\phi H$ in causing this difference.

Several of the enthalpies of mixing in Table 3.4 (see p. 66) have corresponding formation constants listed in Table 3.2 (see pp. 58–66). Taking only the formation constants at comparable temperature and for binary mixtures, the enthalpies of formation of these complexes are determined, and are given in Table 3.5 (see p. 70). These values are compared with directly measured ones in the next section.

Tamres *et al.* [378] and Gordon [152] have shown that there is a roughly linear correlation between the enthalpy of mixing of a donor with trichloromethane, and the acidity constant logarithm, pK_a, of its conjugate acid, for a series of molecules with like donor groups. As Gordon points out, however, it would be more meaningful to correlate pK_a with the Gibbs free energy of association than with $\Delta_m H$; however, $\Delta_\phi G$ data are not yet available for a significant number of trichloromethane complexes.

3.1.4 Enthalpies of Formation

The usual method of determining the thermodynamic properties of an equilibrium reaction is to measure the formation constant, K_ϕ, as a function of temperature. The relation

$$\ln K_\phi = -(\Delta_\phi H/RT) + (\Delta_\phi S/R) \tag{3.9}$$

then yields the enthalpy and entropy changes for the reaction, while the Gibbs free energy change at any temperature is simply

$$\Delta_\phi G = \Delta_\phi H - T\Delta_\phi S \tag{3.10}$$

Unfortunately, this method suffers from the shortcoming discussed in Section 3.1.2—i.e. the reaction whose formation constant is measured involves not simply the formation of a (hydrogen) bond between a molecule of trichloromethane and a donor molecule, but also the breaking and forming of various bonds with solvent molecules as the solvation 'spheres' of the individual molecules are modified to result in the solvation 'sphere' of the complex species. Thus we must consider the reaction scheme [162]

$$
\begin{array}{ccccc}
A(\text{solv}) & + & D(\text{solv}) & \underset{(1)}{\rightleftharpoons} & AD(\text{solv}) \\
\uparrow\,(3) & & \uparrow\,(2) & & \downarrow\,(5) \\
A & + & D & \underset{(4)}{\rightleftharpoons} & AD
\end{array} \tag{3.11}
$$

The enthalpy change measured in solution is the sum of four contributions:

$$\Delta H_1 = \Delta H_4 + \Delta H_2 + \Delta H_3 + \Delta H_5 \tag{3.12}$$

where ΔH_2, ΔH_3 and ΔH_5 are the enthalpy changes for solvation of the three species indicated. Whereas the value of ΔH_4 is desired, it is customary to measure instead ΔH_1.

As may be seen from the enthalpies of complex formation collected in Table 3.6, (see pp. 70–72) most of the values have been measured in solution. Most often, a binary mixture of the two reacting components has been used as the solvent—an obviously unsatisfactory procedure. Not only does the solvent consist of two quite different molecular species, but a change in its composition results in a change in the relative contributions of ΔH_2, ΔH_3 and ΔH_5—of unknown magnitude.

as well as religious. All people in the district, slave and free, white and black, worshipped together, conducted business within their several stations and ranks, ministered and were ministered unto. There was no question of equality among races and classes in the congregation, of course; but the church provided its wide vicinity with a symbol of coherence, of cultural proximity to the center of civilization they looked to, in Charleston, and, beyond Charleston, in the Classical idea of order.

Matthew Mayes, shrewdly but almost certainly out of conviction as well, stayed loyal and active in Salem Black River Church all his life—even while firmly, decisively drawing the *social* focus of the Presbyterian planters away from their historic religious center to the civic settlement he was establishing six miles away. He had already shifted out of the role of urban commercial entrepreneur into that of Jeffersonian rural planter and gentleman, constructed along tidewater lines. Now he began building a role for himself as less the dignified, aloof occupant of an isolated country seat than a leader in a clearly stratified community of planting families with valuable land to protect, and a livelihood to sustain. Like the initial ground scheme of his plantation and future town, the cultural architectonics in Matthew Mayes's mind—how he intended to manage this next transfiguration of himself—has about it an unmistakable sense of *plan.*

That operative strategy was suffused with the traditional notions of the good life—stability, strong social hierarchy, closeness to the land—common to all agrarian visions from Virgil's *Georgics* to Jefferson himself. The first thing to be done was to ground his own identity by successful farming and that of his family as *primus inter pares* in the system of important dynasties dwelling nearby. This he did quickly, assisted by the wealth he had brought to South Carolina, and the wealth and landholdings and local connections he acquired upon his marriage to Henrietta Shaw. Central to this process was the large, serviceable, and marriageable family, which Henrietta gave him—ten sons and daughters, including Junius Alceus, botanist and medical doctor, and Robert, a dry-goods merchant and after 1853 stationmaster at the Mayesville cotton stop. The final and most important step involved binding his properties and children to those of the other leading families, through a system of friendships and business partnerships and marriages—especially marriages among relatives, which is the reason Joe happens to be his own cousin thrice over.

Sallie Burgess's memories of how "he sheltered all in need," and how "many were given in marriage from his fatherly home" ring true, almost. She glides over what Billy Dabbs called the "fierce competition" for farmland that reigned among the Mayesville families. But it was certainly in his role as powerbroker and peacemaker that Matthew became the Squire. And it was his belief in the practical usefulness of blood kinship, woven over time in one place, to keep peace among land-hungry planters that made him the beloved figure of Sallie's memory. From my own standpoint, as a stranger to Mayesville who had sought the Old South there, Matthew Peterson Mayes was the donor to many people scattered across a wide swatch of Carolina ground of perhaps the most precious gift anyone can give another: a sure sense of place.

If *place* has appeared often in this book, with various meanings, I use the word here with the luminously complex weight it carries in the traditional imagination of the South: a dot on the land—a house or farm, at most a village—invested by an awareness of belonging there, and of belonging within the genealogical geometry of ancestors and kin created on that small extent of earth. The designation of anywhere as *homeplace*—an old term in our regional dialect—perhaps best expresses what I am talking about. It is seldom heard now, because few Southerners have a homeplace with which we are uniquely akin. But perhaps *homeplace* was always more noble ideal than historical fact in the Southern mental landscape. Surely, my family has known few such durable dwelling places over its ten generations in America. Our history has been one of uprootings and rerootings, tearings of the fabric of continuity, accumulations of wealth and losses of everything and resurrections, leaving rootlessness as our most durable heritage. There can be no renunciation of this inheritance, which comes from the Modern condition that my English ancestors brought with them to this continent. But in the regret that has come with Modernity—especially in the surge of longing for connection, thick on the ground, that I felt keenly when visiting Mayesville, and for months afterward—I know myself to be a Southerner. In the years before the Civil War, Matthew Peterson

Mayes responded to this pathos of rootlessness in early Republican America by manufacturing, then maintaining all his life, an intricate social fabric in one place, that in turn gave the men and women in his community places to be. Some six generations before I got there, and decades before Sallie Burgess began setting down her thoughts, the historical figure of young Matthew had disappeared into the role of "'the Squire,' as he was lovingly called," in almost the ancient sense submerged in its etymology—the Latin *scutarius*, guard at the court of the Imperator, crafter of shields. This was his gift to the planters on this tiny patch of ground: a complex heraldry, codes stitching together a community and its families—a cohesion maintained, not by force, but by the need to know where one stood on the heraldic field. What was good in the Old South's understanding of place lies not in mansions like those of Gerald O'Hara or Ashley Wilkes, not the hoop skirts and other paraphernalia and memorabilia of modern fantasy, nor in an accent or any artifact, but in the whirring of time's loom, weaving stories and lives on the frame of land into a single swath of meaning. It was this loom I found in Mayesville, and recognized as the creation most to be cherished in all our memories of the South, even if it whirs no longer.

If a homeplace on the ground cannot be saved forever from the batterings of Modernity, the ideal can be recreated, reincarnated, wherever the will to do so endures. For the young Raleigh businessman Matthew Mayes, such an ancient pattern for the good life was available, credible, possible. And it lay ready to hand, in the mind and influence of his elderly Virginia cousin, Thomas Jefferson.

The third President of the republic was always controversial; but during the time I was writing this book, his ideals and deeds had been subjected to more intense critical scrutiny, exposed to more withering attacks and shows of adulation in the scholarly and popular press than at any time since his death in 1826. It is not my intention to add fuel to any fire, still less to presume I have anything new to contribute to the debate. But I have learned from it, and drawn from this dark, crackling air of controversy a certain understanding of the Squire who seems to step from it at times into full daylight, speaking Jefferson's very words, and acting on them. Whether young Matthew ever visited Monticello or the presidential residence in Washington, I do not know. It is entirely possible, given the close entwining of the prominent family of Matthew's mother with her Jefferson kin. But even if his eyes never fell on his eminent kinsman, Matthew's career in South Carolina was nurtured by the loam of Jefferson's dreaming about America, an earthy, agrarian vision destined to wash away by 1861 in all regions of America but the South.

What one is to make of Jefferson's dream is a matter of debate; the contours of the idea itself are not. "Those who labor in the earth are the chosen people of God, if ever He had chosen people," he proclaimed in 1787 in his *Notes on the State of Virginia*, "whose breasts He has made His peculiar deposit for substantial and genuine virtue. It is the focus in which he keeps alive that sacred fire, which otherwise might escape from the face of the earth. Corruption of morals in the mass of cultivators is a phenomenon of which no age nor nation has furnished an example." And, indeed, both the blight of urbanism and the Federalist centralization he opposed would have been banished, in his view, had Americans stayed true to the stubbornly agricultural Roman republic, and in the social virtues that arose there from hard work on the land, and from keeping clear of the infectious stench of industries and cities. This did not mean, however, a retreat into ignorant ruralism. Jefferson was a man too learned, and too politically astute, to despise civilization; he was steeped, after all, in the liberal, elitist Augustan Classicism of his time, and sought to enshrine this ideal of a middle way at Monticello. "Not only is it wrong to envision Jefferson's estates as an integrated agrarian enterprise along the lines of Tara in *Gone With the Wind*," comments the American historian Joseph J. Ellis in one of the most balanced studies to come out of the current squabble,

> not only is it misguided to imagine Jefferson walking behind a plow or spending much time supervising others walking behind a plow, but it is also misleading to think of him residing in a palatial home shaped by his distinctive tastes and filled with his favorite curiosities . . . While his inspirational hymn to the virtuous farmer was unquestionably genuine, the truth was that farming bored him. Retirement to rural solitude did not mean tilling the soil, but digging it up to build something new and useful on it.

The solvents cyclohexane and tetrachloromethane were used in several studies, as well as hexane and ethanenitrile on one occasion each.

The results of greatest interest would be those in the vapour phase, since ΔH_2, ΔH_3 and ΔH_5 are expected to be zero, with the result that $\Delta H_1 = \Delta H_4$ and we have a direct measure of the enthalpy of the hydrogen bond.

One such study of trichloromethane with diethyl ether [130] involved measurement of the second virial coefficient, B_{AD}, for vapour mixtures. The deviation of B_{AD} from a linear dependence on mole fraction leads to the formation constant; this constant, measured at four temperatures, yields $\Delta_\phi H = -25 \cdot 2$ kJ mol^{-1}. A later study of HCCl$_3$ mixtures with several aliphatic ethers has shown that the enthalpy of formation is fairly constant at (-20 ± 5) kJ mol^{-1} [75]. These values are of the same order of magnitude as the enthalpy of one hydrogen bond in benzoic acid dimer [5], and may be taken as consistent with the formation of a hydrogen bond between trichloromethane and ethers. Interestingly, the enthalpy of formation in the binary mixture, ΔH_1, with diethyl ether [326, 250] is almost the same as ΔH_4, implying that $\Delta H_2 + \Delta H_3 \approx \Delta H_5$.

Quite a different situation obtains in the case of complexes with tetra-alkylammonium halide salts. It has been shown that the chemical shifts in solution are consistent with trichloromethane complexes with halide ions [158]. However, the enthalpies of formation in CH$_3$CN solution fall in the range -3 to -6 kJ mol^{-1}. On the other hand, McDaniel *et al.* [103] determined the pressure/composition isotherms for systems consisting of tetraalkylammonium halides (solid salts) and the vapour of HCCl$_3$. Provided that the salt lattice may be assumed to remain unperturbed, enthalpies of formation so measured may be considered to correspond to ΔH_4. The results in Table 3.6 show $\Delta_\phi H$ in the range -35 to -60 kJ mol^{-1}. This implies an enthalpy of complex formation somewhat greater than that of the hydrogen bond in benzoic acid. There remains the complication that $\Delta_\phi H$ is greater for tetrabutylammonium salts than for tetra-ethylammonium ones. The implication is that the smaller cation in the solid salt contributes a greater repulsive energy to the interaction than does the larger tetra-butylammonium. Hence, the measured enthalpy changes should be taken as lower limits to ΔH_4.

The enthalpies of formation in binary mixtures should be comparable to those calculated from enthalpies of mixing and given in Table 3.5 (see p. 70). Comparison shows that the agreement is fair, and probably within experimental uncertainty. In at least some cases, $\Delta_\phi H$ has been measured at high or infinite dilution in one component; hence, the appropriate values of ΔH_2, ΔH_3 and ΔH_5 are not the same as for an equimolar mixture, and are unknown.

The measurements of Macleod and Wilson [250] have apparently been made with great care, and yet are surprisingly inconsistent with the work of others. Their enthalpies of mixing for trichloromethane/diethyl ether are about twice as great as those obtained in other studies (see Table 3.4, pp. 66-69). And yet, they cite a formation constant which yields $\Delta_\phi H$ in agreement with two other studies.

As these authors point out, however, their formation constant leads to a much smaller fraction of combined molecules than found by Dolezalak (see ref 250).

Murakami, Koyama and Fujishiro [276] have found that $\Delta_\phi H$ for the oxolane complex of $HCCl_3$ is considerably more negative than for the oxole and 2-methyloxole complexes. They mention that the electron configuration in oxole is such that the oxygen atom bears a positive charge; hence, trichloromethane will interact preferentially with the π-electrons, this interaction being less energetically favourable than that with the oxygen atom of the saturated ether, and in fact less than that with the π-electrons of aromatic donors.

3.1.5 Effects on the Vibrational Spectrum

Formation of a hydrogen bond has long been known to have a profound influence on the modes of vibration of both the acceptor and donor molecules involved. As Pimentel and McClellan assert in their monograph [309], the vibrational spectrum provides '*the* most sensitive, *the* most characteristic, and *one of the* most informative manifestations of the hydrogen bond'.

As pointed out in Section 2.2, hydrogen bond formation involving the R–A–H molecule results in: (a) a shift of the A–H stretching mode, ν_s, to lower frequency; (b) an increase in the half-width, $w_{1/2}$, of ν_s; (c) an increase in the intensity of ν_s in the infrared spectrum but not in the Raman; (d) a shift of the R–A–H bending mode, ν_b, to higher frequency, with little change in half-width or intensity. Of these, the shift in stretching frequency, $\Delta\nu_s$, is by far the most commonly utilised and reported feature taken to be characteristic of hydrogen bond formation.

3.1.5.1 C–H stretching mode

For many complexes of trichloromethane (and deuterotrichloromethane), the values of $\Delta\nu_1(\Delta\nu_s)$ have been measured, and these are listed in Table 3.7 (see pp. 72-75). It is convenient to record values of $\Delta\nu_1/\nu_1$ so as to place the results for normal and deuterated molecules on a common basis. Where the original workers reported the stretching frequency relative to ν_1 for the vapour phase molecule, $\Delta\nu_1$ is given directly. Wherever trichloromethane in a condensed phase was taken as the reference, the frequency shift has been calculated with respect to the frequency for the vapour phase molecule, and it is this shift which is given in the table. The shift with respect to the uncomplexed molecule in the chosen condensed phase is given in parentheses, and the nature of the condensed phase given in parentheses after the solvent.

The other columns give the values of half-width of the stretching band, its intensity relative to that of the stretching band of the uncomplexed molecule, in the cases where these values have been measured.

As Pimentel and McClellan pointed out [309, p. 197], the usual criterion of specific interaction, which is $\Delta\nu_s$, provides ambiguous evidence of hydrogen bonding of trichloromethane. With oxygen-donor bases, $\Delta\nu_s$ is often about the same as in non-donor solvents. Nitrogen-donor bases, however, and halide-ion donors produce substantially larger shifts of the stretching frequency. Nevertheless, the oxygen donors, and indeed the π-donors also, lead to considerable increases in the intensity of the ν_1 band in the infrared spectrum, although not in the Raman spectrum [322]; Pimentel and McClellan consider this increase to be adequate confirmation of the presence of hydrogen bonding, in spite of the absence of the expected shift. Curiously, the band half-width increases in the case of oxygen donors but does not for π-donors. The intensity of ν_1 was also found to increase with concentration in tetrachloromethane or cyclohexane— an indication of self-association [27].

In view of the situation for alkyne acceptors (see Section 5.2.5), it is pertinent to note that there is no systematic difference of $\Delta\nu_1/\nu_1$ between the ordinary and deuterated molecules of trichloromethane [106].

In Bishui, Mukherjee and Sirkar's low-temperature Raman study of trichloromethane [39], they find a small but irregular decrease of ν_1 in the solid phase with respect to the liquid-phase value. This, and small changes in the other vibrational frequencies, are attributed in no very convincing way to the formation of weak intermolecular hydrogen bonds.

The shift in ν_1 of $DCCl_3$ solutions containing $[(C_4H_9)_4N][(C_6H_5)_3PMI_3]$, where M = Ni(II), Co(II) and Zn(II), has been taken [324] to signify interaction of trichloromethane with the iodide portion of the anion ($DCCl_3$ solutions of triphenylphosphine show no such effect). The same shift in all three cases argues that the nature of the interaction is in each case the same, although different from that of $(C_4H_9)_4NI$ which produces a much larger change in ν_1. Intensity measurements show that approximately $4 \cdot 5 \pm 0 \cdot 5$ molecules of trichloromethane interact with one molecule of $[(C_4H_9)_4N][(C_6H_5)_3PZnI_3]$ or of $(C_4H_9)_4NI$. It was also found that triphenylphosphine and triphenylmethane each interact with two to three molecules of trichloromethane, presumably via the aromatic rings.

The intensity of ν_1 has been used as a measure of the 'strengths' of hydrogen bonds involving the trichloromethane molecule. Using this criterion, Hanson and Bouck [167] find the following phosphine oxide molecules to be ranked thus:

$[(CH_3)_2N]_3PO$ strongest
$(C_6H_5O)_3PO$
$(C_2H_5S)_3PO$
$(C_6H_5S)_3PO$
$(4-CH_3C_6H_4O)_3PO$
$(3-CH_3C_6H_4O)_3PO$
$(2-CH_3C_6H_4O)_3PO$
$[(C_2H_5)_2N]_3PO$ weakest

They offer no explanation for the placing of the two nitrogen-containing molecules at the two extremes of the series.

The intensity of ν_1 is found to be smaller in the halogenated ketones $CFCl_2-CO-CFCl_2$, $CFCl_2-CO-CF_2Cl$, and $CF_2Cl-CO-CF_2Cl$ than in pure trichloromethane [202]. This is probably due to a smaller degree of association in these solvents than in the pure liquid—a sharp contrast to the case with unsubstituted propanone.

3.1.5.2 *C–H bending mode*

Little evidence concerning the bending mode frequency, ν_4, is available for complexes of trichloromethane. Several groups of workers have reported values of $\Delta\nu_4$; their results are given in Table 3.8 (see p. 76). It is interesting that, in the presence of non-donating or weakly donating bases, the bending frequency shifts to *lower* values whereas the presence of donating bases causes shifts to higher frequencies.

The results for $HCCl_3$ and $DCCl_3$ seem to be substantially different, whether complexed with weak bases or strong. Part, at least, of the discrepancy may be due to the fact that the magnitudes of the shifts being considered are quite small; hence, small experimental errors become magnified when reported in this way. Nevertheless, particularly with the strong donors, this explanation is insufficient. It appears that, indeed, $(\Delta\nu_4/\nu_4)_{CD} < (\Delta\nu_4/\nu_4)_{CH}$.

Pimentel and McClellan [309] comment that, characteristically, $\Delta\nu_b/\nu_b$ is considerably smaller than $\Delta\nu_s/\nu_s$, and that $w_{1/2}$ for the bending mode is unaffected by hydrogen bond formation. A comparison of the results in Table 3.8 with those in Table 3.7 shows that these generalisations are not applicable to complexes of trichloromethane. The complexes with nitromethane, diethyl ether, di(2-propyl) ether, propanone and dimethyl sulphoxide—all oxygen-donor bases—exhibit $(\Delta\nu_b/\nu_b) > (\Delta\nu_s/\nu_s)$, while the nitrogen donor bases have the expected $(\Delta\nu_b/\nu_b) < (\Delta\nu_s/\nu_s)$.

As found by Szczepaniak [372], the half-width of the bending mode increases by a factor 2 to 3 in the presence of strong donors. This feature is presumably characteristic of complex formation involving trichloromethane.

The bending mode peak due to a complex with propanenitrile or ethane-nitrile is found to increase in intensity at lower temperatures at the expense of the peak for unassociated $HCCl_3$ [268]. However, equimolar mixtures which are frozen at 93 K show a considerably greater perturbation of ν_4 in the case of propanenitrile than with ethanenitrile [268]. It seems evident that propanenitrile and trichloromethane form 1:1 complexes exclusively, in a solid solution; on the other hand, no complex exists between $HCCl_3$ and ethanenitrile, presumably reflecting the difference in proton acceptor abilities of the two nitriles. The molecules of ethanenitrile in a solid mixture with trichloromethane are present in two different environments, in one of which the crystallisation is influenced by the $HCCl_3$.

Cotton field, near Mayesville

But America, in Jefferson's view, should nevertheless stay as close to the soil, and to the tilling of it, as possible. Ellis comments:

> Domestic manufacturing was permissible, but large factories should be resisted. Most important, the English model of a thoroughly commercial and industrial society in which the economy was dominated by merchants, bankers and industrialists should be avoided at all costs. "We may exclude them from our territory," he warned, "as we do persons afflicted with disease. . . ."

Jefferson did ease up on this vigilant stricture eventually, and wanted his revision to be made public. "We must now place the manufacturer by the side of the agriculturist," he wrote to Benjamin Austin in 1816. Those who could not see the necessity for small manufacture "must be for reducing us either to dependence on that foreign nation"—meaning England—"or to be clothed in skins, or to live like wild beasts in dens and caverns."

That being said, the essential role of the Jeffersonian planter remained the same: a rural example of the median struck between the extremes of antique refinement and the barbarism of Modernity. But an example for whom? The social model clearly assumes a continuing differentiation between great farmers and small ones, preceptors and receptors; and the evolution of this class-divided culture within simple, unhurried country life. Its inscription is perhaps most readily traceable in the best Southern Classical architecture of Jeffersonian inspiration—in the present context, for instance, the magnificent "Brick Church," with its muscular Doric columns and strong, refined entablature, raised in 1846 for the Presbyterians of Salem Black River. Such Classical temples "growing in the woods of America," as he put it, should express the simple morality of early Rome, the liberty of the Greeks; and do so, to this day. But such values are also embedded in the self-limiting character of the Squire's understanding of Mayesville. Unlike other antebellum county towns in the Old South, and unlike every town in the New South, Mayesville was founded small with the prospect of remaining small. Matthew Mayes's plan was, it appears, not to set in motion a juggernaut of expansion, spreading out remorselessly from a business and governmental center, but rather to establish a firm point in an ecological system of civilized services, worship, and sociability, and farming under the sun.

The contradiction at the heart of Jefferson's conservatively Modern notion of liberty—the fact that such freedom to plant and think could only be bought by slavery—is the primary target of the present-day argument over his intellectual and moral legacy. It is also an ineradicable torment in the heart of anyone who has ever longed, as I have, to celebrate freely the Old South of cultivation and connection so richly manifested in the tapestry of lives and destinies woven at Mayesville. And it is this bitter conflict that makes barren and hollow the consolation Catherine Deveraux Edmonston, mistress of Looking Glass Plantation, North Carolina—one of the legion of intelligent Southern women who have left behind eloquent journals of the war years—gave herself in 1863. "Though my young countrymen may not be so skilled in the learning of Greece and Rome, and the quadrature of the circle and the mysteries of the conic section may be a sealed book to them," she wrote in her diary,

> let us trust they bring back from the battlefield a knowledge of men, of the secret springs of the human heart; and fitted, as they will be by having learned obedience, to govern, the future of our country in their hands will be both glorious and prosperous. That War, while she strips from them many a modern refinement, and the wisdom of schools, will gild them with her own barbaric virtues—a lofty contempt of danger, a chivalric devotion to women, a spirit of self-sacrifice which will make them spring to the defense of the weak, a devotion to their country, a love of honor which shall be their guiding star through life.

Nor did the mind of Squire Mayes escape the South's tragic hope of glory so eloquently incarnate in Catherine Edmonston's words. For in trying to do so—in joining the attempt to keep one section of American soil safe for the exercise of doomed Jeffersonian ideals, and for the construction of an agrarian culture kept in place by the joining of families and the mingling of blood, and based on the preservation of a race of slaves—he put what Sallie calls "his beautiful signature" on South Carolina's Ordinance of Secession. In doing that, he condemned a son to death on the battlefield, and helped consign his state, and all the South, to immeasurable suffering in both war and postwar peace. And he exposed whatever was decent, and whatever was of encouragement to the human spirit in the pastoral, anti-Modern version of the republic conceived by his great cousin, to merciless reprisal, and the scorn from which it would never—could never—recover, except in the dreams of the defeated, and in the memories of those who will not forget the profound good in that rejected version and vision of the republic.

Several workers [153, 154, 345] have observed that mixtures of trichloromethane with an electron donor give rise to a fairly intense absorption near
2450 cm^{-1}; Thompson and Pimentel [381] have interpreted this as due to the
first overtone of ν_4, which is much more intense than the fundamental. The
frequency of this bond with several solvents is given here:

Solvent	$2\nu_4/cm^{-1}$
tetrachloromethane	2401 ± 2
propanone	2448 ± 5
diethyl ether	2457 ± 5
triethylamine	2504 ± 5

The high intensity of this overtone is suggested by Thompson and Pimentel
[381] to be due to the fact that it depends on charge mobility along the
hydrogen bond, whereas the intensity of the fundamental, ν_4, is proportional
to the charge mobility perpendicular to that bond.

The intensities of ν_1 and ν_4 show a marked dependence on the isotope of
hydrogen, the former being greater in HCCl$_3$ by a factor of about 2, the latter
being greater in DCCl$_3$ by a factor of about 3. Borisenko and Shchepkin [46]
have calculated that $\delta\mu/\delta r_{CH(D)}$ and $\delta\mu/\delta r_{CCl}$ do not show appreciable differences
between the two isotopic forms of trichloromethane. Hence, the *proportion* of
each contributing to each observed mode must be responsible for the observed
differences, a change in proportion of $\delta\mu/\delta r_{CCl}$ being particularly effective in
altering intensities.

Devaure, Turrell, Huong and Lascombe [106] have measured *all* the vibrational
frequencies of the molecules HCCl$_3$ and DCCl$_3$ in the vapour phase and in a
number of solvents. The modes ν_1 and ν_4 are simply the C–H (or C–D) stretching
and bending modes, respectively. The low frequency mode, ν_6, could only be
measured in hexane, trichloromethane and triethylamine solutions; it has the
value 260–261 cm^{-1} in all cases. In addition, ν_3 is observed to be essentially
independent of the molecular surroundings, as well as being nearly the same for
the two isotopic forms of trichloromethane.

On the other hand, ν_2, the symmetrical stretching vibration of the CCl$_3$ group,
is markedly dependent upon isotopic substitution, and also on the nature of the
solvent. It would seem that coupling of the C–H bond with the motion of this
group is responsible for both observed effects, since the vibrations of the C–H
group are affected by both. The frequencies are given in Table 3.9 (see p. 77).

The vibrational mode ν_5 exhibits similar behaviour on isotopic substitution
and on change of solvent [106]. In this case, $\Delta\nu_5/\nu_5$ differs for the two isotopic
forms from one solvent to another. Devaure *et al.* postulate that this phenomenon
is due to a sort of coupling between the various vibrations of type E. Then the
rule

$$\sum_i \Delta\nu_i/\nu_i = \sum_i \Delta\nu_i^*/\nu_i^*, \quad i = 4, 5, 6 \tag{3.13}$$

B

is expected to hold, where the quantities indicated by an asterisk denote values for DCCl$_3$. They find that this rule is indeed obeyed. A similar isotope- and solvent-dependence of ν_5 (and ν_4) has been found by Borisenko and Shchepkin [46].

Oi and Coetzee [290] have found the C–Cl stretching frequencies of several chlorinated hydrocarbons to be dependent upon solvent. They find for 1-chloro- and 2-chloropropane that these frequencies are linearly related to the solvent parameter $(\epsilon_r - 1)/(2\epsilon_r + 1)$, ϵ_r being the solvent permittivity relative to the

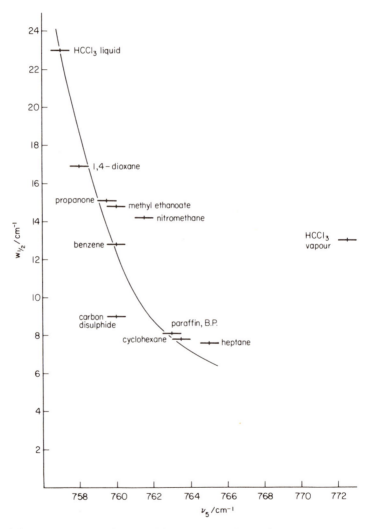

Fig. 3.1 A comparison of the linewidth and wavenumber of the vibrational mode ν_5 of trichloromethane in several solvents.

PART THREE

PILLARS OF CLOUD AND FIRE

Library
I.U.P.
diana, Pd.

permittivity of a vacuum. The solvents benzene and 1,4-dioxane produce larger-than-predicted shifts owing to local polarity and polarisability which are not reflected in the bulk permittivity of these solvents.

However, ν_2 and ν_5 of trichloromethane show additional deviations from linearity in the solvents triethylamine, ethyl ethanoate, propanone, ethanenitrile and nitromethane. This behaviour is undoubtedly due to complex formation, particularly so since ν_2 exhibits separate bands in propanone, triethylamine and diethyl ether due to vibrations of different species. This multiplicity of C–Cl stretching modes has also been observed by other workers [158, 106].

Whiffen [398] has noted a similar solvent variation for the value of ν_5 in HCCl$_3$. In addition, he finds that the linewidth of this band increases as its frequency decreases—i.e. as the solvent donor ability increases. He asserts that 'no numerical correlation with any particular property of the solvent readily presents itself', although Figure 3.1 suggests that the same factor may be influencing the frequency changes.

3.1.5.3 Effects on the donor molecule

The vibrational modes of donor molecules are affected by hydrogen bond formation, though usually to a lesser extent that those of the acceptor [104]. For example, it was found [153] that the C=O band of propanone or ethyl ethanoate near 1760 cm^{-1} is shifted to lower frequency on dissolving in trichloromethane. Similarly, the C–O band of diethyl ether or 1,4-dioxane, at 1240 cm^{-1} and 875 cm^{-1}, respectively, are displaced to lower frequency in trichloromethane solution.

Lascombe *et al.* have compiled the stretching frequencies of the carbonyl group of a number of ketones upon complex formation with trichloromethane in hexane or cyclohexane solvent [83; see also 372 and 205]. The differences of these frequencies from those of the uncomplexed molecules are given in Table 3.10. In each case, three separate bands are observed in the 1620 to 1750 cm^{-1} region, and their intensities vary with the relative amounts of the two components. The authors consider these observations to be firm evidence for the existence of complexes of types II and III shown in Section 3.1.1. Complexes of type IV would not be likely to result in a separate band for the C=O stretching mode.

The Raman frequency of the S=O stretching mode of dimethyl sulphoxide is found to be 1054 cm^{-1} in CCl$_4$ solution and 1054–1051 cm^{-1} in trichloromethane, a negligible difference [313]. This is in marked contrast to the behaviour of the infrared-active band which is observed at 1071 cm^{-1} in tetrachloromethane but at 1055 cm^{-1} in HCCl$_3$ [30]. These two bands have recently been shown [306] to be due to 1:1 and 2:1 complexes of trichloromethane with dimethyl sulphoxide, while a band at 1080 cm^{-1} is attributed to uncomplexed dimethyl sulphoxide.

The P–O stretching frequency of a number of compounds is slightly lower in trichloromethane solvent than in an inert solvent [163, 167].

The C≡N stretching frequency of benzonitriles shows a slight *increase* on changing solvent from CCl_4 to $HCCl_3$ [380]. The intensity of this band, however, increases very markedly on changing to the more strongly interacting solvent, an effect which Thompson and Steel attribute [380] to hydrogen-bond formation perturbing the dipolar nature of the C≡N bond.

3.1.6 Effects on the NMR Spectrum

3.1.6.1 Chemical shifts in the complexes

Table 3.11 (see pp. 78–87) contains a listing of the changes in chemical shift of the proton in trichloromethane on changing from one medium to another. In most cases, the second medium is, or contains, a donor species which may be expected to form a hydrogen bonded complex with $HCCl_3$. The trichloromethane hydrogen resonance moves to lower field, as would be expected on formation of a hydrogen bond with a donor species. In some aromatic solvents, the complex is formed with the π electrons of the donor molecule. The large magnetic anisotropy of the aromatic donor results in a net shift to higher field [94, 323].

Table 3.12 gives several values of Δ_s as predicted from the data of Tables 3.2 (see pp. 58–66) and 3.11 (see Section 2.3). These are seen to be in reasonable agreement with the observed solvent shifts in these solvents. Hence the simple model of two $HCCl_3$ environments as presented in Section 2.3 appears to be a reasonable one.

Recently, Wiley and Miller have reported the results of an unusually thorough and careful study by NMR spectroscopy of trichloromethane complexes with a number of electron donors [400]. They affirm the opinion that the proton shift of $HCCl_3$ in a pure solvent, and hence the commonly reported solvent shifts and dilution shifts, is not a very meaningful quantity.

A further aspect of Wiley and Miller's work is concerned with the temperature dependence of the complex shift of the trichloromethane proton. They find that the magnitude of Δ_c decreases as the temperature rises, and ascribe this to a weakening of the hydrogen bond through vibrational excitation in a shallow, anharmonic potential 'well' (i.e. $-\Delta H$ is ca. 4 to 17 kJ mol^{-1}), in accord with the prediction of Muller and Reiter [273]. However, the situation is not clear, since $\delta(\Delta_c)/\delta T$ should decrease as $\Delta_\phi H$ becomes larger; but the few examples illustrated in reference [400] show the opposite behaviour, i.e. $\delta(\Delta_c)/\delta T$ becomes larger as $\Delta_\phi H$ increases.

This observation may be compared with that of Green and Martin [158] who found that the complex shift of $HCCl_3$ with halide ions is invariant with temperature, even though the value of $-\Delta_\phi H$ obtained from the temperature dependence of the formation constant is only about 4–5 kJ mol^{-1}. It is evident in this case that the values of ΔH_4 and ΔH_1 are considerably different, the former being more reasonably given by the gas–solid phase value of 45–60 kJ mol^{-1} [103].

Hence, Wiley and Miller's findings are a source of puzzlement, since they seem to show that, for example, the dissociation enthalpy of the $HCCl_3$/triethylamine complex is smaller than that of $HCCl_3$/thiolane.

Rettig and Drago [324] find that the proton chemical shift of solvent trichloromethane is observed at successively higher field as portions of the salt $[(C_4H_9)_4N][(C_6H_5)_3PNiI_3]$ or $[(C_4H_9)_4N][(C_6H_5)_3PCoI_3]$ are added. They attribute this unusual effect to a contact shift, indicating that unpaired electron spin is transmitted to the $HCCl_3$ molecules through a partially covalent link.

Friedrich has pointed out [134, 136] that the magnitude of the solvent shift of trichloromethane in amine solvents with respect to hexane solvent is much affected by the steric crowding about the nitrogen donor atom. He shows that the solvent shift varies monotonically with the Taft E_s parameters, which

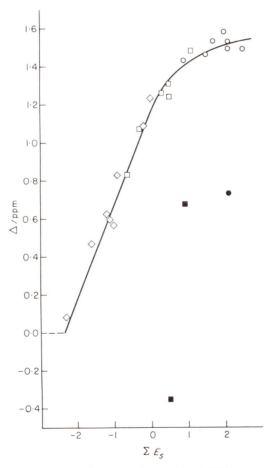

Fig. 3.2 Relationship between the ^1H NMR solvent shift of trichloromethane (with respect to hexane) and the Taft steric factors, E_s, of the substituents on the amine solvent molecule, $NR_1R_2R_3$. Tertiary: ◇; secondary: □; primary: ○; aromatic: filled symbol.

THE ASHES OF ILIUM

Burned building, Shreveport

give a measure of the steric requirements of the alkyl groups, although the steric factor becomes relatively unimportant when $\Sigma E_s > 1$, and the solvent shifts of the primary amines fall within quite a narrow range (see Fig. 3.2). When the substituent(s) on the amine are not saturated hydrocarbons, additional factors such as polar effects also affect the trichloromethane solvent shift and the points for such solvents do not fall on the curve of Figure 3.2. The filled symbols on the figure are for amines with aromatic (phenyl- or phenylmethyl-) substituents; these illustrate the effects of (a) the magnetic anisotropy of the substituent on the solvent molecule, and (b) more important, the effect of competing equilibria to form N-donor complexes and π-donor complexes with trichloromethane.

Friedrich shows, further, that, although the solvent-induced shifts in tetramethylsilane and benzene are proportional to the solvent volume magnetic susceptibility for a wide variety of solvents, such a correlation is not valid when trichloromethane is the solute molecule [136]. Compared with hydrocarbon solvents, the $HCCl_3$ signal appears at lower field than that expected on the basis of the relationship with susceptibility. However, within each class of solvents (hydrocarbons, chlorinated hydrocarbons, nitriles, ethers, alcohols, ketones and amines) the solvent-induced shifts obey a linear relationship with χ_v, except where steric crowding is a factor. Hence, there is a contribution to the solvent-induced shift of $HCCl_3$ from hydrogen-bond formation in addition to the susceptibility contribution. The hydrogen bonding contribution appears to depend only on the nature of the donor group in the solvent molecule and not on the nature of the remainder of the molecule (except for steric effects). These hydrogen bond contributions are as follows:

Group	$>C-Cl$	$-C\equiv N$	$>O$	H $>O$	$>C=O$	$>N$
Hydrogen Bond Shift/ppm	0·22	0·53	0·70	0·87	0·90	1·53

Friedrich has shown [135, 136] that these hydrogen bond shift contributions are directly proportional to the atomic orbital dipole moments of the donor atoms, except for nitriles (see Fig. 3.3). The value for nitriles, when corrected for the effect of the anisotropy of the $C\equiv N$ group, is also found to be proportional to the atomic orbital dipole moment.

Effects on the donor spectrum. Laszlo has shown [234] that the proton chemical shifts exhibited by many molecules when dissolved in trichloromethane are somewhat different from the shifts in tetrachloromethane solution. The signals of donor molecules were found to appear at lower field in almost all the cases examined. Presumably the complex (hydrogen-bond) formation results

OVER THE FIRST 250 YEARS OF MY TRIBE'S HISTORY IN AMERica—between 1609, when the Reverend William arrived at Kecoughtan, and the early 1860s—the Mays families in America sustained no calamitous shock to the way they lived, or the paths of thinking they walked. To be sure, these men and women never tilled the same plot of Southern ground for more than two or three generations before loading up their wagons, harnessing their mules, and moving on. But such upending and translation was the usual experience of all early Americans, their customary culture, and could be accommodated into a sensible picture of the world's unfolding.

The coming of cotton certainly wrought radical transfigurations on the economy of the South, changing patterns of land use, and transforming the enslavement of blacks and aborigines from a traditional practice into a crucial institution, buttressed by new ruthlessness, and by the variously paternalistic and philosophical ideologies created to justify it. But if my colonial Virginia ancestors had thought tending tobacco and hogs with a few slaves to be the proper occupation of a gentleman, their sons and grandsons along the Saluda River, and their cousins in lower South Carolina, quickly found cotton an acceptable agricultural substitute. If a novel crop, hard on the earth and hard to make profitable without much-enlarged plantations, cotton wrought little discord in the old harmonies of my people with the soil and seasons, and allowed them to go on living much as English folk had always lived in America since their first spring plantings along the rivers of seventeenth-century Virginia.

Then came the war, bringing with it the stark and subtle transmutations in the South, and in the life of countless families, mine among them.

John Matthew Mays, my great-grandfather, was born an heir to our region's settled rural traditions in 1846, on the Saluda plantation of his people. He was fifteen years old when his cousin, Matthew Peterson Mays, and many other prominent, passionate men put their signatures to the Southern states' Ordinances of Secession, pro-

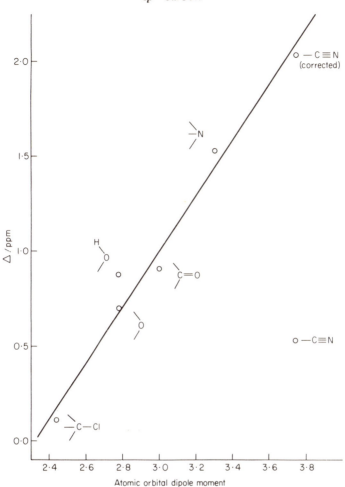

Fig. 3.3 Hydrogen bond shifts of HCCl₃, Δ, due to certain electron donor groups, as a function of the atomic orbital dipole moments of those groups. Reprinted from *Zeitschrift für Naturforschung*, Vol. **20b**, 1965, p. 1030. Copyright 1965 by H. J. Friedrich. Reprinted by permission of the copyright owner.

in a small reduction of electron density in the donor molecule and therefore a reduction in the shielding of its nuclei.

3.1.6.2 Coupling constants in the complexes

The one-bond $^{13}C-^{1}H$ coupling constant in trichloromethane has been determined in a variety of solvents [116, 391]; the values are given in Table 3.13 (see p. 88). Evans argues that the mutual electronic repulsion between the donor electrons on the base and the C—H bonding electrons leads to an increase in the 's' character of the bonding orbital [116]. Since the magnitude of the carbon–hydrogen coupling

pelling the Old South, and all the good and evil we had created there, on its march toward crisis, disaster, and the transfiguration of the Southern mind.

John was serving his temporary nation as a scout on the Savannah River—watching through the chilly spring mist for the lights of Union gunships the dying Confederacy could not have stopped—in the dawn hours of Monday, April 3, 1865, when the last Rebel commanders at Richmond ordered all gas lamps in home and street extinguished, the munitions dump set on fire, and the capital's center torched. The victors entered the blazing Confederate capital just after sun-up; and, among the fires and half-blinded by smoke, at 8:15 that morning, representatives of the United States accepted Richmond's formal surrender. President Lincoln arrived from Washington on Thursday, sat for a symbolic moment at the desk from which Jefferson Davis had directed the states in revolt, then toured the smoldering ruins of our Southern Troy.

Judith McGuire, refugee from western Virginia hosteled at Richmond, wife of an Episcopalian priest and Confederate officer, his fate still unknown to her, had watched as the streets of her haven were darkened and its buildings set alight. Later, from her spot among the appalled throngs gathered on Richmond's sidewalks, she saw the first laughing Union troops emerge from the smoke and bitter drift of cinders, and the stooped American President pass slowly, mindfully, through the wreckage—she watched, then withdrew into her rooms behind shutters, waiting and writing.

Though she was no kinswoman of mine, Judith's raw, telegraphic journals preserve invaluable records of what women across the South, including those in my family, felt and endured in those early, uncertain days of defeat. At first, Judith's entries are fitful and fearful, darkened by anticipation of unnamed horrors yet to come. But these frightened jottings disappear from her chronicle as she finds the occupying Yankee soldiers, white and black, to be more gentlemanly and better-mannered than expected. As the victorious Union troops settle in, the looting and rape of Richmond she feared do not materialize. And when they do not, the frightened Virginian in Judith McGuire retires from her writing, and the veiled, solemn writer returns, reflecting on Richmond's self-inflicted fires and the ruin they left, and on the fires of Southern patriotism, now snuffed out. Only with General Lee's surrender of the Army of Northern Virginia to Grant a fortnight later does Judith realize the only world she had ever known is vanishing, becoming a topic unworthy of writing's conserving and transmission. Toward the end of her diary she notes: "How all this happened—how Grant's hundreds of thousands overcame our little band, history, not I, must tell my children's children."

But even after Lee's departure from the field, Judith continues to believe that resistance elsewhere has not been wholly crushed. Then comes the news of General Johnston's surrender to Sherman of all men still under Confederate arms between the Potomac and the Rio Grande. Judith jots down the facts about the end without apparent emotion, making no attempt to interpret them. At the end of this dry accounting, she appends five words abruptly charged with the infinite poignancy and finality of a parent's whisper to a dying child: "My native land, good night."

And with that, the troubled night began to fall over Judith McGuire's Southern homeland—immediately on the exhausted forces in Lee's rebel Army of Northern Virginia and Grant's loyal Army of the Potomac; and quickly on the Southern cities and command posts to which the news of defeat was brought at once by newspapers and telegraph and military messenger. The end came more slowly to the river overlook where young John Mays kept watch, and slowest to the remote hollows and fastnesses of the mountain South, where Confederate soldiers, hungry and isolated and desperate, went on fighting the enemy and their own weariness. Now fleeing west and south, without a country or armies to defend it, the tatters of the Confederate government issued its last proclamations into the failing light, urging "a new phase of struggle," and pledging its faith in "fresh defiance" they believed would spring from the warriors' "unconquered and unconquerable hearts."

A sizable number of men who heard acknowledged their antique, chivalric version of American destiny to be doomed in the territory of the industrial, urban United States, and opted for desertion from the reunited country. The eccentric General Joseph Orville Shelby, for instance, marched his solders into Mexico and put them at the disposal of the Emperor Maximilian, then fighting off desert guerrillas, only to see his Confederates go over to the rebel side. Sobered

constant is strongly dependent on the 's' character of the bonding orbital [272], Evans' argument is consistent with the observed values of $^1J_{CH}$ (see also Section 3.2.7).

Lichter and Roberts [242] have measured the solvent shifts, Δ_s, of the ^{13}C signal of $HCCl_3$ in the same solvents; these are also given in Table 3.13. They find a linear correlation between Δ_s and $^1J_{CH}$, but regard this as perhaps fortuitous since the solvent shifts are small, and carbon-13 shifts are easily influenced by a variety of factors.

A curious feature noticed by Lichter and Roberts is the fact that the carbon-13 chemical shifts of $HCCl_3$ in benzene correlates well with J_{CH}, while the proton shift is well known to appear at higher field because of the magnetic anisotropy of the benzene ring. This observation may imply that, although the hydrogen atom of trichloromethane is situated on or near the axis of the benzene ring in the complex species, the carbon atom is not. More likely, the ^{13}C chemical shifts are dominated by factors other than the anisotropy of the solvent molecule.

3.1.6.3 Complex relaxation times

By measuring the spin-lattice relaxation time, T_1, of the protons of several donor molecules in solution, Anderson [11] has been able to show whether the donor molecule rotates as an independent entity, or as a part of a larger complex molecule with the solvent. In the solvents benzene, tetrachloromethane, carbon disulphide and trichloromethane, the solutes benzene and propanone are found to rotate as independent molecules, even though they are known to be associated with $DCCl_3$ to a considerable extent. The solute molecules 1,4-dioxane, dimethyl sulphoxide and 2-methyl-2-aminopropane all rotate more slowly in $DCCl_3$ than in the other solvents, showing that it is a larger, complex molecule whose rotation is being measured in these cases. The same conclusion regarding trichloromethane/1,4-dioxane mixtures was reached from a study of the broadening of the anisotropic Rayleigh scattering component [13a], which showed that the relaxation time of reorientation for the mixtures is considerably greater than that of either component.

Anderson postulates that the molecules of benzene and propanone, even though forming a complex with trichloromethane, undergo rotation independently by means of large-angle jumps between minima in a potential energy surface, rather than by a continuous, diffusion process.

3.1.7 Miscellaneous

3.1.7.1 Dipole moments of the complexes

Measured values of the dipole moments of complex compounds of trichloromethane are given in Table 3.14 (see p. 188). It is evident that these values are not very reproducible from one group of workers to another.

As Jumper and Howard showed [201] for the complex with diethyl ether, a change of solvent from cyclohexane to tetrachloromethane results in an increase of ca. 1.5×10^{-30} C m; it appears that solvent molecules are polarised in the vicinity of the complex, adding to the measured moment. Such a consideration renders suspect all the values obtained under conditions other than high dilution in a hydrocarbon or other inert solvent. Thus, the trichloromethane/triethylamine complex has a moment of 6.90×10^{-30} C m in cyclohexane, but 10.5×10^{-30} C m in the binary mixture.

In the studies of complex moments, the aim has been to demonstrate the charge-transfer nature of the hydrogen bond involving $HCCl_3$. It is apparent that the results obtained to date are not conclusive in resolving this question. One difficulty is the changes in the dipole moments of the donor and acceptor molecules on forming a complex. Because the polarisation interaction has the effect of changing the dipole moments of the molecules, we do not know their moments as they occur in the complex. This factor renders impossible the aim of using the complex dipole moments to ascertain the complex geometries.

In every case, the difference between the additive molecular moments and the measured complex moments is small. This small difference is consistent with the theoretical and experimental finding that only with fairly strongly acidic acceptors is there an appreciable change ($\geq 1 \times 10^{-30}$ C m) in the moment upon complex formation [321]. Hence, the degree of charge transfer in complexes of $HCCl_3$ is small.

A study by Małecki, using dielectric polarisation and saturation measurements was aimed at discovering the composition of the complex(es) formed in trichloromethane solutions of dodecanol-1 [254]. He found that the postulate of either a 1:2 or a 2:1 complex yields a calculated complex moment in close agreement with the experimental value. Thus, although the stoichiometry of the complex is not known unambiguously, nevertheless it is clear that the additive moments of the components agree with the complex moment.

3.1.7.2 *Dielectric constant and loss*

Antony and Smyth have made a number of dielectric constant and loss measurements on binary mixtures of trichloromethane [13]. They find that the reduced relaxation times of $HCCl_3$, τ_0/η, are a function of solvent, as shown here.

Solvent	$(\tau_0/\eta)/10^{-14}$ Pa^{-1}
1,4-dioxane	13
benzene	11.0
trichloromethane	9.6
tetrachloromethane	5.1
cyclohexane	3.3

and disillusioned, Shelby soon returned, and ended his days as a Missouri farmer. More desperate, durably romantic souls—in all, some three thousand, according to one estimate—signed up with the Southern Colonization Society, led by my South Carolina kinsman Major Joseph Abney, and other organizations sponsoring exodus and self-exile. Joseph's dream was to imitate the refugee Aeneas, and sail forth with his troops from our ruined Troy to reconstruct an Old South in, of all places, the jungles of Brazil. He and his followers and comrades were fortunate; the Brazilian Emperor Dom Pedro II gave them free land, freedom to practice their Methodist faith and to own slaves, and to be as Southern they liked. More than a century later, thousands of *los Confederados* can still be found in southeastern Brazil, practicing Methodism and speaking Portuguese—now thoroughly mixed with the local populations, but still holding meetings of the *Fraternidade Descendencia Americana* to tell stories and keep alive the history of their remarkable emigration.

But by far the greater number of boys and men in gray made shorter journeys, and quick truces with the new reality. One was John Matthew Mays. We do not know exactly when John heard the war was over, and headed home. From the diaries of upcountry women, newspaper reports, and family memories, however, we catch glimpses of the grim Carolina landscape John traversed on his long walk upriver, back to the plantations of his ancestors and kin. The fields along the Saluda had largely escaped direct devastation by Union troops. But the region had not escaped, any more than the rest of the South, the demoralization, and fear of famine and disease and general misery of spirit that fell on my countrymen in the spring and summer of 1865. Throughout the hill country back of all our coasts, unsubdued Rebel platoons broke off from their legions, changing from warriors into looters and terrorists, as defeated men have always done, when abruptly loosed from the disciplines of authorized killing and left stranded, *without place,* on burned, strange ground.

John Mays knew where his place lay, and defied danger and hunger to reach it, only to find his plantation near Cokesbury wasted by neglect, the freed slaves gone, fences decayed, family treasures unrecoverable—his mother had dropped the silver into a well to save it, thereby losing it forever—and mother, stepfather, and younger

The authors suggest that the relaxation time of an electron acceptor should increase with electron donor ability of solvents. This order is borne out by the relaxation times of trichloromethane.

In mixtures with 1,4-diazabicyclo[2.2.2]octane, two relaxation times of $HCCl_3$ are found. A reduced relaxation time of $7 \times 10^{-14} Pa^{-1}$ is appropriate for uncomplexed $HCCl_3$ molecules. However, the other reduced relaxation time, $39 \times 10^{-14} Pa^{-1}$, is attributed to the rotation of a trichloromethane/donor complex. No such distribution of relaxation times was found for the donor solvent 1,4-dioxane.

The complex dielectric constants of trichloromethane/azine mixtures at several temperatures yield relaxation times which exhibit clear maxima for equimolar ratios. Savchenko, Vakalov and Shakhparonov assert [334] that these observations are consistent with the formation of complexes between the two components.

3.1.7.3 Magnetic susceptibilities

Gopalakrishna has reported the excess magnetic susceptibilities of several binary mixtures of trichloromethane [149, 150, 151]. The attempt to interpret negative excess magnetic susceptibilities in terms of a reduction of diamagnetism is not convincing in view of both internal inconsistencies (see text and figure of [151]), and inconsistencies with other work (e.g. hydrogen-bond formation with N of aminobenzene, whereas the complex forms with the π electrons [283]). Magnetic susceptibilities would appear to offer results of doubtful reliability and hence to be of little usefulness.

3.1.7.4 Azeotropic mixtures

It was found by Segal and Jonassen [346] that trichloromethane forms a maximum-boiling azeotropic mixture with ethylamine [$x(HCCl_3) = 0.57$ mol mol^{-1}; b.p. 341·4 K]. Presumably because of the large difference between the boiling points of $HCCl_3$ and diethylamine, these substances are found not to form such a mixture [346].

The latter phenomenon was observed also by Ewell and Welch [119] who found that $HCCl_3$ does not form azeotropic mixtures with diethyl ether (b.p. 308 K), 2-methyloxirane (b.p. 308 K), triethylamine (b.p. 362 K), or 1,4-dioxane (b.p. 374 K), even though fairly strong interactions with all of these are known to be present. Donor molecules with boiling points closer to that of trichloromethane do form maximum boiling azeotropic mixtures: dimethoxymethane (b.p. 315 K), propanal (b.p. 323 K), 2,2-dimethyloxirane (b.p. 323 K), ethyl methanoate (b.p. 327 K), propanone (b.p. 329 K), methyl ethanoate (b.p. 332 K), 2-methylpropanal (b.p. 336 K), di(2-propyl) ether (b.p. 341 K), butanal (b.p. 349 K), and butanone (b.p. 353 K). The azeotrope of trichloromethane and

sister reduced to near-destitution by the collapse of the South's wartime economy. John's job was to survive; and into his own survival and that of his kin, John threw all the energies he possessed, hoeing the weedy earth, putting in enough food crops to keep alive his family and what cattle remained after that spring season of broken fences and thieving by desperate former comrades.

In the homeplaces of my other relations scattered across South Carolina, the same heavy desolation of spirit and body prevailed. At Mayesville, the old Squire, in the cemetery of the Brick Church, mourned by the grave of his son William, who, at age twenty-four, had died hideously of gangrene following battlefield surgery. Other plantation families in his district waited for their offspring to arrive, never certain whether they would return at all, or whether they would come back living or dead. "The community was torn, a way of life was disrupted," writes local historian Ethel Cooper Turner, "and everyone faced a period of hard and trying times ahead during [R]econstruction." Trains arrived at the station seldom, she continues, arrivals and departures at the station were uncertain. The carriages hauled only slight supplies because they were so heavily loaded with demobilized Confederate soldiers, seeking home.

Here, as everywhere in the South, the train brought weary, hungry sons or boxes to bury, and the waiting stopped. But not all the children of the Squire's town reappeared. A tale still told in Mayesville recalls a woman who watched the locomotives come, stop, and go throughout the summer of 1865—trains that did not bring her offspring back that year, or in the years afterward; who, all her life, kept vigil on the front verandah of her house, gazing down the road in expectation of a son destined never to return.

Another of John's relations, Samuel Elias Mays—grandson of the old warrior Samuel Mays—fought the last days of war with soldiers "hungry and almost naked, and some of the men . . . actually barefooted," as he tells us in a memoir, then went home to Pendleton, in upcountry South Carolina, where he had practiced law before the war. Like my other cousins and kin who had fought for the Confederacy, he came back to vanished certainties. Rich before the war, Samuel found that his $80,000 investment in Confederate bonds, and some $20,000 received for a wartime sale of property—in Confederate currency—were gone, and he was left holding a fortune

di(2-propyl) ether has a composition of 0·38 weight fraction HCCl$_3$ and a boiling point of 343·4 K.

3.1.7.5 Vapour pressures

Findlay and Kenyon [126] have measured the vapour pressures of azine/trichloromethane mixtures at 323·1$_5$ and 336·6$_5$ K. They find consistently negative deviations from the values calculated assuming the validity of Raoult's Law. Such a result is invariably observed when the *inter*-component interaction is stronger than the intracomponent interactions. It is interesting to note that the deviations from 'ideal' behaviour are much greater at the higher temperature; the inference must be that the difference in strengths of interaction is greater at higher temperatures.

3.1.7.6 Volume of mixing

The excess volume of mixing, V^E, of trichloromethane/azine mixtures at 298·2 K was found by Findlay and Kenyon to be small and negative [126], with the minimum values at a trichloromethane mole fraction of 0·68. Similar results [364] for trichloromethane/propanone mixtures would seem to indicate stable complexes involving two molecules of trichloromethane in each case. The geometry of such a complex with azine would be problematical.

The values of V^E for trichloromethane mixtures with dialkyl ethers, on the other hand, are large and negative [24], whereas mixtures of CCl$_4$ with the same ethers exhibit much less negative values of V^E. By contrast, V^E for HCCl$_3$ mixtures of dimethyl sulphoxide is about as negative as for CCl$_4$ mixtures with the sulphoxide [71, 236]. Chareyron and Cléchet account for this behaviour [71] by concluding that the interaction is strong between dimethyl sulphoxide and tetrachloromethane!

3.1.7.7 Electronic spectrum

Balasubramaniam and Rao have observed that the n → π* transitions of diphenyl-ketone and 1,3-dithiolane-2-thione undergo a marked shift to higher frequency in trichloromethane solvent relative to heptane solvent [17]. They assert that these shifts, comparable in magnitude to those in alcohol solvents, are too large to be accounted for by change in permittivity of the solvent, and so must be due to solvent–solute hydrogen bonding. Perhaps it is significant that much smaller frequency shifts were observed for other donor solutes in the same pair of solvents (i.e. heptane and trichloromethane).

The energies of interaction which these workers attribute to hydrogen bonding, and which are listed in Table 3.6 (see p. 70), all lie between 1·3 and 7·5 kJ mol^{-1}, values which are very small—of the order of RT. In addition, the spectra illustrated do not exhibit a clear isosbestic point (1,3-dithiolane-2-thione/trichloromethane),

"with the value about the same as that much waste paper." After walking the deserted streets of Charleston at war's end, he wrote:

> I could not help recall to mind a speech I heard old Mr. Maverick make at Pendleton, S.C., in 1850, at the time when secession was beginning to take tangible form. Said the old gentleman in allusion to it: "My friends, you could do nothing more foolish. Your ships will rot at your wharves, or they will desert your cities. Your commerce will be destroyed and grass will grow in the streets of Charleston." Now I have seen all this come to pass.

In the summer of 1865, the time had come for Southerners to bury our dead, to welcome home the living to devastation and tears, and, for a few rare souls like Samuel Elias Mays, to think back upon the prophetic warnings that the men of my family, and of myriad other families, did not heed.

The time had not come, however, for the *other* kind of remembering—wreathed in nostalgia, drenched with longing for lost innocence—that was to take root and flourish, like a sweet, toxic flower, in the soil of postwar Southern imagination.

It was just a seedling in the South during the first, severe years of Reconstruction, this myth of the Lost Cause. But it was to be nourished steadily by the telling and retelling of tales by men and women growing older, ever more distant from the terrible events. Over the years of our region's political Reconstruction and economic recovery, and far into the era of New South prosperity that would be crushed by the Great Depression, the skirmishes in these narratives gradually became battles, and a little raiding expanded into a spectacular assault on enemy positions. The stories of mere survival came to be embroidered with acts of daring and comradely heroism, and the unspeakable warfare wreathed with glory. Many a man who fought the Yankees, and especially men who died doing so, underwent an apotheosis to some higher degree of magnificence than his war record would have supported, had anybody bothered to look. My Great-Grandfather John, the river scout, for example, came to be known as "Captain Mays," a title he never held in his short military career, and a veteran of several important battles he never saw.

I am certainly not alleging that John Matthew Mays, or any other honorable survivor of the Civil War, set out deliberately to deceive anyone. What is at work in this fabulation is not a willful failure of individual memory, but the ancient, sociable alchemy of mythmaking by the defeated—the process of gradual exaltation, by unlegislated consensus and assent, of historical fact into transcendental example, as a way to ease and dignify the pain of remembering. The Greek and Roman historians, poets, and moralists taught Southerners how such transformation is done, and so did the prophets of ancient Israel, as they sought to wrest some elevating meaning from the destruction of their highest dreams.

The problem for many Southerners of my generation is that our ancestors were almost *too* successful at displacing the inner experience of Civil War with shining, prophetic myth in the decades following surrender. To be sure, the passionate rhetoric of the "Redeemer Nation that died," but "survived as a sacred presence, a holy ghost haunting the spirits and actions of post-Civil War Southerners," as cultural historian Charles Reagan Wilson has summarized it, was a spent force by the time I was born, without authority or even much interest. Almost no trace of it remained in the small mental world my closest kin and I inhabited during my earliest years of growing up. Neither Sister Erin nor Aunt Vandalia ever said anything about the war I can remember—except, on the rare occasions the topic came up, that the Confederacy had been in the wrong, slavery was evil, the racial segregation that succeeded slavery was right, and the South's defeat had been the will of Almighty God.

And in the primary schools I attended, glorification of the Lost Cause, and even much mention of the Cause itself, had been dropped from the curriculum. Our civics and American history textbooks had been crafted according to the official Cold War ideology of the United States, which emphasized the unity of all Americans under the Constitution, and treated the South's rebellion as an unfortunate glitch in the operatic unfolding of the nation's imperial destiny. Our teachers did not even take us to Civil War battlegrounds, though a couple of minor ones lay close by. When I eventually visited such sites on my own, I found empty pastures, and sun-faded charts under Plexiglas, decorated with what looked like schools of dolphins twist-

so no complex species is clearly assignable. The evidence from this study seems to be inconclusive.

Kuboyama has presented the spectrum of acenaphthenequinone in several solvents [227]. He attributes a weak band near 500 nm to an n → π* transition and asserts that a shift to higher frequency in trichloromethane solvent is evidence of a hydrogen-bonding interaction, while a similar shift in 1,4-dioxane solvent is due to molecular complex formation. He suggests no means of differentiating between the two types of interaction.

Becker, likewise [28], attributes the high-frequency shift of the n → π* transition of diphenylketone in trichloromethane solvent (767 cm^{-1} with respect to hydrocarbon solvent) to hydrogen bonding. The slightly smaller shifts in RCH_2CN solvents (507 cm^{-1}) were attributed to dipolar complex formation.

3.1.7.8 Polarisation

Ke and Lin have reported a study of trichloromethane complexes employing polarisation measurements at optical frequency [210]. Strangely, they chose ethanol as the 'indifferent' solvent for this study. Considering the well-known hydrogen bonding properties of this solvent, any attempt to measure weak hydrogen bonding equilibria in it must be regarded with strong suspicion.

3.1.7.9 Compressibilities of mixtures

Parshad has measured ultrasonic velocities in trichloromethane mixtures with diethyl ether and with propanone [295]. He finds the compressibility of the diethyl ether mixtures to be less than expected on the basis of additivity, whereas that of the propanone mixtures is greater than expected. He argues that these observations may be taken as indicative of dipole–dipole interaction with the ether, and hydrogen bonding with the ketone. Similar measurements by Venkatasubramanian [388] also showed that mixtures with propanone show large deviations from the expected behaviour.

As Parshad correctly points out, most measurements of the interactions of trichloromethane with donor molecules do not distinguish between hydrogen bonding and other types of interaction. He draws the distinction that hydrogen bond formation involves the (partial) transfer of an electron pair to trichloromethane, but concludes 'that there is really no difference of kind but only of degree between hydrogen bonds and dipole association' [295].

3.1.7.10 Ultrasonic absorption coefficient

The mole fraction dependence of the ultrasonic absorption coefficient in mixtures of non-associating liquids is found by Sette to be concave upward [349]. The mixture trichloromethane/benzene exhibits 'anomalous' behaviour in that the concentration dependence is linear. Analogous results were obtained for propanone/methanol mixtures. This behaviour is taken to be evidence of attractive interaction between the components, but seems rather unconvincing.

3.1.7.11 *Refractive index*

Segal and Jonassen [346] have observed that the refractive indices of trichloromethane/ethylamine mixtures exhibit a discontinuity at about equimolar concentrations, which they take as possibly indicative of an attractive interaction between the components.

3.1.7.12 *Freezing point diagrams*

These have been determined for the binary systems of trichloromethane with ethanenitrile, propanenitrile, butanenitrile and benzonitrile [277]. The last three indicate the presence of 1:1 molecular addition compounds, while the butanenitrile also forms a compound with three molecules of trichloromethane, $C_3H_7CN.3HCCl_3$.

Interestingly, the $HCCl_3/CH_3CN$ diagram shows no evidence of formation of a molecular compound. Murray and Schneider [277] speculate that the self-association of ethanenitrile is so strong that the trichloromethane is unable to compete successfully in forming a mixed compound.

Wyatt has found [403] a number of molecular compounds from freezing point diagrams of trichloromethane mixtures, as follows: $(C_2H_5)_2O.2HCCl_3$; $(C_2H_5)_2O.HCCl_3$; $2(C_2H_5)_2O.HCCl_3$; $3(C_2H_5)_2O.HCCl_3$; $(CH_3)_2CO.HCCl_3$ [404]. No compound was found in benzene/trichloromethane mixtures.

He also found [403, 405] that tetrachloromethane exhibits molecular compound formation with diethyl ether, propanone, *and benzene.* The implication would seem to be that the chlorine atoms are more important to complex formation than is the hydrogen atom.

3.1.7.13 *Second virial coefficient*

The second virial coefficient of a mixture of two vapours may be expressed as

$$B = (1-x)^2 B_{11} + x^2 B_{22} + 2x(1-x)B_{12} \qquad (3.18)$$

where x is the mole fraction of substance 2. B_{11} and B_{22} are the values of the second virial coefficient for the pure vapours 1 and 2 respectively, while B_{12} is characteristic of the mixture, and represents the intermolecular forces between 1 and 2.

In a mixture of substances whose molecular force fields are similar, it might be expected that B_{12} would be nearly the arithmetic mean of B_{11} and B_{22}, in which case a graph of B vs. composition would be linear. Fox and Lambert [130] have found that, for trichloromethane/hexane and diethyl ether/hexane mixtures,

ing and turning in formation, indicating charges and maneuvers. It meant nothing to me.

But in my family, however, there was the old story about our ancestor who had dropped her silver down the well. Around the tale, whenever my grandmother or Aunt Vandalia told it, hung a certain cloud of melancholy, of undefined loss—a suggestive fragrance rather than an explicit statement, suggesting that we Southerners were like that silver, shining in the murk of a time now dominated by rude, aggressive Yankees, whom we must never emulate. If never spelled out that clearly in my hearing, this exemplary lesson settled into the depths of my mind, where it helped feed the obsession with the South that began growing in me around age twenty.

It was at this time—after my first reading of the *Iliad* and the Nordic sagas, and bewitched by Poe—that I wrote the first of a number of darkly gothic Civil War tales. One told of a childless woman alone on her dead warrior-husband's coastal plantation; upon hearing the thunder of the Union rabble approaching, she sets fire to the great house over her head, offering her body as a final sacrifice to the shade of her fallen spouse. As she casts a torch on the bier of furniture her slaves have heaped in the magnificent hall, she invokes the fury of the South Carolina sea winds—which "hit the pyre, and a huge inhuman blaze rose, roaring. / Nightlong they piled the flames on the funeral pyre together / and blew with a screaming blast . . . " (The quote, from Book XXIII of the *Iliad*.)

At least one spur to my slow, painful abandonment of this ill romanticism, during the mid-course of the civil rights struggle, was the revival of Lost Cause rhetoric in the service of racism. One heard no note of nostalgic romance in the blather of George Wallace, governor of Alabama, who was, to my mind, a no-neck monster risen from the swamp of lesser life forms. He and those who spoke like him were the folk of shotguns and pickup trucks, racist swagger and bullying, who wore Rebel flags on their baseball caps—not the sort of people, that is, to honor the name of Robert E. Lee, whom every Southerner of my class and generation was taught, at home in any case, to revere as the finest flower of Southern Christendom. I despised them, for reasons having less to do with conviction than with the bias of my caste, but found I could not disentangle what they said from the Southern goodness I was trying, in my curious manner, to preserve.

B is linearly related to composition. However, the B vs. composition graph is concave downward for trichloromethane/diethyl ether mixtures. This behaviour is consistent with the formation of a loose intermolecular compound; the data may be used to derive values of the equilibrium constant for its formation, and hence the enthalpy change involved. These values are reported in Tables 3.2 (pp. 58-66) and 3.6 (p. 70). Similarly, B_{12} for dimethyl sulphoxide mixtures with trichloromethane is found to be smaller than the arithmetic mean of B_{11} and B_{22} [308].

The second virial coefficient of pure trichloromethane vapour was found by Lambert *et al.* [233] to follow closely the temperature dependence predicted by the Berthelot equation (from 320 to 395 K). This is regarded as evidence for the absence of long-lived dimeric molecules in the vapour phase at these temperatures (although there is some deviation at the lower temperatures). The dipole–dipole interaction of a pair of $HCCl_3$ molecules is calculated by the authors as -2300 J mol^{-1}, which is less than the thermal energy RT.

3.1.7.14 Solubilities

Marvel, Dietz and Copley [261] point out that the solubility of donor molecules is likely to be higher in trichloromethane than in tetrachloromethane, because of the opportunity for hydrogen bond formation with the former solvent. They find the ratio of solubility in the two solvents is always greater than unity, and can be as high as 10^2.

These authors measured the solubilities of a number of polymers in several chlorinated solvents at 298 K. They find the following ordering of solubilities [261]:

$$HCCl_3 \gtrsim HCCl_2CHCl_2 > HCCl{=}CCl_2 > CCl_4 > CCl_2{=}CCl_2$$

This ordering is attributed to the effects of different hydrogen-bonding abilities.

3.1.7.15 Viscosity

Macleod has shown [249] that it is possible to deduce, from the measured viscosities of binary mixtures, information about the interaction of the components. He emphasised that the viscosity and density data used must be of a very high order of accuracy. At 273 K, he concludes that about 20% of the molecules in a trichloromethane/diethyl ether mixture have combined to form complex molecules.

3.2 The Other Trihalomethanes, HCX$_3$

The complex formation properties of the trihalomethanes other than $HCCl_3$ have been studied to a lesser extent and hence fewer quantitative data are available concerning their complexes. In the progression from low to high atomic

My disillusionment was hastened by the coincidence of civil rights with the centenary of the Rebellion's defeat, both of which made the papers and television news often in the 1960s. If civil rights was the issue that counted, the Civil War was the topic that sold; and volume after volume about the "glory" or the "tragedy" or the *whatever* of the Civil War littered the bookshops in my university town. Going back over my journals of that period now, decades later, trying to recall what attention I paid to the centenary—in common with all diarists, I write things down so I can forget them—I have unearthed only a few curt comments about some fancy-dress reenactment at a Civil War battlefield in Pennsylvania that I thought ridiculous. And since that time, the fascination that inspires Americans to stitch together uniforms of the Grays and Blues, confect battle flags, and fire off puffs of gunpowder at one another in the suffocating heat of our Southern summers for the sake of tourists, has continued to seem perhaps less silly to me than merely incomprehensible.

It is quite possible that, after the 1960s and my own attachment to the Old South were done, I might have gone on the rest of my life without a thought about the Civil War had I not gone looking for the traces left by my ancestors and kin in the war's aftermath. Indeed, I almost bypassed Richmond on my Virginia travels between coastal Hampton and the Staunton, undertaken early in the writing of this book. If every Southern town has its sculptural tribute to the men who died for the Cause, the capital of Virginia is wholly such a monument—papered thick with plaques and bits of statuary erected to honor the fighters of our wretchedly mistaken war, and to commemorate the politicians who made their postwar reputations defending its rightness. (In my view, Richmond should be stripped of all these trappings, and completely redecorated as a shrine to Poe, the city's greatest adopted son and the most influential literary mind ever molded and inspired by Southern culture.)

Despite these inner objections, however, I did eventually drop by Richmond one day, mostly because I wanted to see Thomas Jefferson's small Classical masterpiece, the state capitol building. It was almost by accident—driving around, trying to find a place to park— that I ended up at the Museum of the Confederacy, which stands next to President Davis's White House. I vaguely expected something repellent. What I found was intriguing—a treasury of portraits, flags,

number, the molecular dipole moment decreases monotonically and the polarisability increases similarly. Hence, a similar ability to engage in complex formation may be expected throughout the series.

	Dipole moment $\mu/10^{-30}$ C m	Polarisability $\alpha/$C m^2 V^{-1}
HCF_3	5·5	
$HCCl_3$	3·7	$9·16 \times 10^{-40}$
$HCBr_3$	3·3	$13·16 \times 10^{-40}$
HCl_3	2·7	$19·7 \ \times 10^{-40}$

3.2.1 Stoichiometry of the Complexes

As is found to be the case for trichloromethane, the other trihalomethanes form mostly 1:1 complexes with donor molecules. Unlike trichloromethane, many of the higher complexes which have been detected are of stoichiometry AB_3 (see Table 3.15; cf. Table 3.1).* It has been shown, in the studies carried out by X-ray determination of crystal structures, that the interaction is between the donor atom of the base molecule and the *halogen* (iodine) atom of the trihalomethane molecule [116, 180].

This complex geometry leads to the conclusion that the interaction is of a polarisation or charge-transfer type. A hydrogen bond type of interaction is presumably less energetically favourable than the interaction involving all three halogen atoms. Even the A_2B complex $2HCl_3$.1,3-diselenane, in which one molecule of triiodomethane interacts with each atom of selenium, involves interaction of the selenium with an iodine atom.

The existence of trihalomethane complexes involving bonding at hydrogen or at halogen argues that the two forms of interaction must be of comparable energy. This aspect is explored further in Chapter 6

3.2.2 Complex Formation Constants

The values of the trihalomethane complex formation constants are given in Table 3.16. As with trichloromethane, these formation constants are dependent on both temperature and solvent, as may be seen for the tribromomethane complexes with tetrabutylammonium halide salts. In addition, the reporting of results in different units of concentration renders comparisons difficult.

Nevertheless, it is possible again to compare the extent of complex formation, in the case of tribromomethane, between halide ions and covalent donor atoms. In common with trichloromethane, $HCBr_3$ is complexed more extensively with the halide ions than with donors such as azine, oxygen donors and tribromomethane itself.

* See p. 89; cf. p. 57.

weapons, photographs, and other artifacts sharply evoking the brief era of our independence from the United States. The museum, as it turned out, was neither morgue nor sanctuary nor shrine, but a repository of precisely the hard, often cruel and arresting facts obscured in the rhapsodizing of Lost Cause myth and racist oratory.

Toward the end of my tour of the museum, I discovered one of them—the tale of the C.S.S. *Shenandoah,* commemorated there by heraldry and a few souvenirs. For many months after the surrender, as it happened, the deck of this armed cruiser, pushing through the icy waters of the northern Pacific, remained the only unconquered territory of the Confederacy. And, as I later discovered—in a report that still rings true despite its deployment of *Shenandoah* for Lost Cause ends—this tiny patch of planking was never to be conquered by the Union or by anyone else.

The teak-hulled ship had been dispatched in February 1865, at Melbourne, Australia, with instructions from Richmond to hunt down Yankee ships whaling in the Arctic. Captain James Iredell Waddell and his crew rounded the Kamchatka Peninsula in early June, entering the Bering Sea, and soon began meting out destruction to the enemy's commercial fleet. It was soon after their arrival in that Arctic summer's endless day—the account comes from the *Shenandoah*'s executive officer William C. Whittle—that the Confederate destroyer captured the *William Thompson* and the *Susan Abigail,* both out of San Francisco, and Captain Waddell found on board one of his captive ships an April newspaper. A story in its pages related the final breakdown of the Army of Northern Virginia at Appomattox. Another "contained President Davis's proclamation from Danville, VA, stating that the surrender would only cause the prosecution of the War with renewed vigor."

The officers of the *Shenandoah* decided to ignore the bad news and follow their President's directive, hunting down, bombarding, and burning Yankee whalers for the next several weeks. In August, reported Whittle, the crew turned southward and caught sight of a fuming volcano—"the last land which we were destined to see for a long time." And so it was to be: for off the coast of California they captured a craft Waddell believed to be American, but turned out to be British, and carrying the news Captain Waddell had feared since June: "the surrender of the Confederate forces, the capture of President Davis and the entire collapse of the Confederate cause." Whittle continues: "Coming as it did from an Englishman, we could not doubt its accuracy. We were bereft of country, bereft of government, bereft of ground for hope or aspiration, bereft of a cause for which to struggle and suffer. The pouring of hirelings from the outside world had at last overpowered the remaining gallant Confederates."

All on board the *Shenandoah* knew the ship would now be hunted ruthlessly by American cruisers. Whereupon Captain Waddell made up his mind never to surrender to the Yankees, and to find safe harbor in a land far from the Union. With the colors of the Confederacy downed, the *Shenandoah* set off again south, rounding Cape Horn, and, after several months of steaming north across the Atlantic Ocean, finally putting in at Liverpool. After this "long, weary and anxious voyage, with its share of gales and storms," wrote Whittle, the cruiser was handed over to the British, who later sold it to the Sultan of Zanzibar, whereafter it disappeared from history. The crew returned to North America, where most of them, too, disappeared from sight, leaving only a few men to write annals of what they believed to have been an epic voyage of freedom.

"Noble man! chivalrous soul! brave heart!" exclaimed Captain J. A. Ashe in ecstatic apostrophe to Captain Waddell before a meeting of the Ladies' Memorial Association, in Raleigh, on May 10, 1902.

> We here after these many years behold you rising aloft in those distant waters, the sole and solitary Confederate banner that has floated upon the bosom of the ocean. Alone it is borne by the breeze over the great waste of waters—the only emblem of our nation's sovereignty beyond the limits of our beleaguered states. We now realize the difficulties that beset you. We know the perils of the deep—the storms and hurricanes that sweep the ocean—the fury of the wild waves moved by mighty winds—but these, these have no place in your thoughts as you unfold the flag of your country, then heroically struggling for existence, but your mind is intent only on the honor of your countrymen!

By 1902, we are reminded by Captain Ashe's encomium to the absent commander of the *Shenandoah*, the enormous ideological indus-

3.2.3 Enthalpies of Mixing

The enthalpy changes on mixing trihalomethanes with an equimolar amount of a second component are given in Table 3.17 (see p. 92). As is the case for trichloromethane, it is not self-evident how these are related to the formation constants and enthalpies of formation of the respective complexes. All the enthalpies of mixing listed here are somewhat smaller than the corresponding values for $HCCl_3$. Indeed, the tribromomethane values of $\Delta_m H$ are only about half as great as are those for $HCCl_3$. Perhaps the smaller dipole moment of tribromomethane results in a smaller formation constant, or enthalpy of formation, or both. Unfortunately, neither quantity has been measured in $HCBr_3$ complexes with aromatic molecules.

3.2.4 Enthalpies of Formation

The values of enthalpies of formation of trihalomethane complexes are given in Table 3.18. As compared with trichloromethane, the data are rather sparse; this is to be regretted, since the series of trihalomethanes provides a systematically varying set of properties, such as dipole moment and polarisability, which could usefully serve to demonstrate the importance of the various characteristics of the acceptor molecule in complex formation.

Again, as in Section 3.1.4, almost all the reported values are complicated by the fact of their being measures of ΔH_1, where the enthalpies of solvation of the various species are unknown.

3.2.5 Solubilities

Zellhoefer, Copley and Marvel [406] have measured the solubility of fluorodichloromethane, $HCCl_2F$, in a large number of solvents of various types. Their results are listed in Table 3.19 (see pp. 93–95). They find that: (a) ether oxygen produces high solubility; (b) sulphur is a much less effective donor than oxygen; (c) an ester group donates electrons about as well as an ether oxygen atom; (d) carbonyl oxygen is also a good donor group; (e) alcohols are fairly poor donors because of extensive self-association; (f) an amide group is a more effective donor than an ester; (g) covalent fluorine is not an effective donor group. All of these conclusions are based on the assumption that the solubility of fluorodichloromethane depends on its electron accepting properties.

The same authors compared the solubilities of a number of substituted methanes in several oxygenated solvents. They found that the enhancement of solubility over the 'ideal' value exhibits the following trends:

$$H_3CCl < H_2CCl_2 < HCCl_3$$
$$H_2CFCl < H_2CCl_2$$
$$HCClF_2 < HCCl_2F < HCCl_3$$

3.2.6 Effects on the Vibrational Spectrum

3.2.6.1 The acceptor molecule

The changes in the C–H stretching mode frequency, ν_s, of the various trihalo-methanes upon complex formation are given in Table 3.20 (see p. 95). Unlike trichloromethane, vapour-phase values of ν_s are not available for all of these compounds. Consequently, it is not possible to express the frequencies in the complex species with respect to those for the respective free (vapour-phase) molecules.

Nevertheless, it is apparent that the fractional shifts $\Delta\nu_s/\nu_s$ are a maximum for complexes of tribromomethane, becoming smaller as lighter or heavier halogens are substituted for bromine.

It is dramatically apparent that substituents such as the halogens are necessary for such large changes in stretching frequency to be observed, when compared with the behaviour of the alkyl-substituted analogue methylpropane, $(CH_3)_3CD$. The gas-phase stretching frequency of the C–D bond is 2152 cm^{-1} [105]. The same mode in a variety of solvents including hexane, dibromomethane, 1,3,5-trimethylbenzene, ethanenitrile and triethylamine, is observed to fall within the narrow range 2143–2148 cm^{-1}.

3.2.6.2 The donor molecule

As is the case for trichloromethane solvent, the C=O stretching frequency of ethyl ethanamide is lower in HCBr$_3$ solvent than in CCl$_4$ or CS$_2$ (1663 cm^{-1} vs. 1687 cm^{-1}) [218]. Also, the S=O stretching frequency of dimethyl sulphoxide is shifted to 1070 cm^{-1} and to 1050 cm^{-1} by HCBr$_3$, these values being attributed to 1:1 and 2:1 complexes by Philippe and Cléchet [306].

3.2.7 Effects on the NMR Spectrum

As may be seen from the data of Table 3.21 (see pp. 96–98), the measured chemical shifts on complex formation of the trihalomethanes are of two types–solvent shifts and complex shifts (see Section 2.3). In general, for both types of shift, we observe that, with a common donor solvent, the magnitudes of the shift differences fall in the progression HCCl$_3$ > HCBr$_3$ > HCl$_3$. The data for trifluoromethane suggest that its shifts fall between those of HCCl$_3$ and HCBr$_3$, but are too few for this to be certain.

The complex shifts of trihalomethanes with halide salts have been inter-preted [158] in terms of the Buckingham model [55] of chemical shifts in the X–H bond. This model led to the conclusion that imposing an electric field on an X–H bond results in a change in chemical shift given by

$$\Delta\sigma = -AE_z - BE^2 \tag{3.19}$$

where E_z is the field strength along the bond axis, E^2 is the square of the field strength, and *A* and *B* are constants characteristic of the atom X. Hence, for complexes of the same geometry with the same donor, all the trihalomethanes are predicted to experience the same change in shift. The differences in observed shifts then imply either (i) a variation of *A* and/or *B* with the substituents on X, or (ii) a variation in complex geometry.

The differences in complex shifts were interpreted [158] as primarily due to the second factor. The shifts to *higher* field observed for triiodomethane complexes were taken to show that these complexes involve interaction of the *iodine* atoms of the acceptor rather than of the hydrogen. It was further deemed probable that a halide ion interacts with all three iodine atoms simultaneously, rather than with only one. This model is consistent with the complex shifts which are of smaller magnitude and opposite sign to those of the other trihalomethanes.

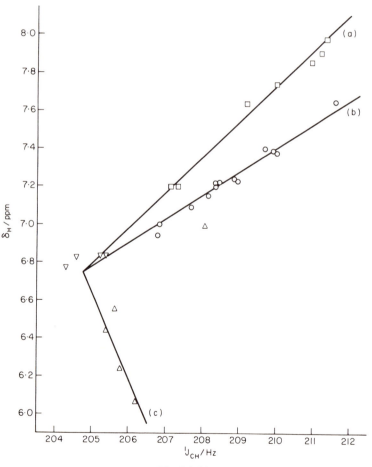

Fig. 3.4 (a)

try of the Lost Cause, manufacturing formulaic tributes and imagery and tropes and inscriptions beyond numbering, was running at high speed. The South's first monument to General Stonewall Jackson had been dedicated some twenty-five years before in what an old soldier called "memory-fraught Richmond, the soldier's Mecca"; and two hundred thousand were to gather in the same city in 1907 for the unveiling of a cenotaph raised to the memory of Jefferson Davis. And everywhere across the South, veterans and their families and friends were putting up memorials of the Confederate Dead, none more famously imprinted on the world's vision than the slender soldier on the square at Oxford.

The first American reporters on the scene, in April 1865, at Appomattox Courthouse—perhaps better journalists than I—knew they had a good story, and played it large. In the earliest newspaper dispatches from Virginia, readers were given portraits of the arrogant Ulysses S. Grant suddenly humbled in triumph, admitting to his sadness "at the downfall of a foe who had fought so long and valiantly, and had suffered so much for a cause. . . ." Hearing of unseemly celebrations that had broken out in the Union states after the Southern enemy's destruction, General Grant is said to have declared: "The war is over, the rebels are our countrymen again, and the best sign of rejoicing after the victory will be to abstain from all demonstrations." The tidewater aristocrat General Lee was never more noble than in his mass-media portrayal. In an address to his wretched soldiers in the Army of Northern Virginia, he commands them to turn their hands to "the allayment of passion, the dissipation of prejudice, and the restoration of reason," and to take up their role voluntarily as citizens in the rejoined nation (that they, of course, and Lee himself, had so energetically helped rip apart).

And the popular historians and military enthusiasts who have since walked a thousand times in these reporters' tracks to the final fields of war have always known the power of a good story, and inflated their accounts accordingly. Faithful to such melodramatic iconography of the event, the twentieth-century historian James M. McPherson writes of the famous meeting of Lee and Grant in the living room of Wilmer McLean:

> The vanquished commander, six feet tall and erect in bearing, arrived in full-dress uniform with sash and jeweled sword; the

victor, five feet eight with stooped shoulders, appeared in his usual private's blouse with mud-spattered trousers tucked into muddy boots. . . . There in McLean's parlor the son of an Ohio tanner dictated surrender terms to the scion of a First Family of Virginia.

McPherson's breathless description is academic history-painting in words, replete with every mechanism such a vast moral, allegorical picture is supposed to have: dramatic chiaroscuro, compelling realism, compositional sweep, the radiances and glooms of classical tragedy. In his use of such colors, allusions, and tones, McPherson merely reproduces the principal motifs in the first published accounts, which themselves were murky mixes of much fancy and some fact. The result is banality on a large canvas.

But like other history-painting, the imagery McPherson and others created to memorialize the events of April 9, 1865, cannot be read as a factual representation of the past, whatever we may mean by the word *fact*, or *past*. It is writing that simplifies, *decontaminates* the matter, and makes it quickly digestible for contemporary consumers, by setting it within an orderly and recognizable moral frame. In fairness to this dubious project, we should perhaps recall that the writers were almost certainly trying to answer profound human desires for intellectual order, giving myth to people whose sensitivities had been so deeply wounded that they had become impervious to fact. Any contemporary obituary of one who is said to have "fought heroically" against this or that disease reminds us that fitting disaster into a literary framework remains as central to existence at the end of the twentieth century as it was in the nineteenth. A terrible outcome must have a good reason, or at least a good moral lesson. It does not matter, of course, whether those in such need have ever read or heard of Homer or the other great literary enframers of disaster. The longing for epic explanation is written on our hearts.

At least one thing elevates such morally profound accounts of war as the *Iliad* and the *Aeneid* (and *The Red Badge of Courage*, and a heartbreaking poem about battlefield defeat by Miss Dickinson), and made them crucial reading in the time I was trying to understand the aftermath of civil war. It is the readiness of such great ancient and modern poets in the West to show us the horror of battle *as* horror—

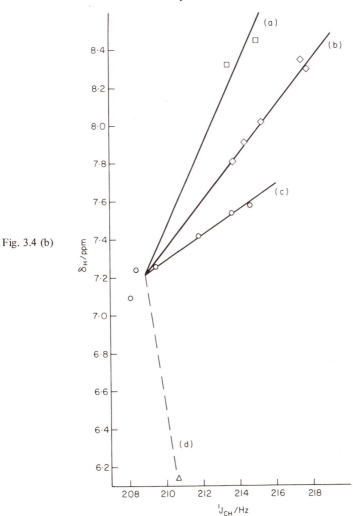

Fig. 3.4 (b)

Fig. 3.4 (a) The relationship between ^{1}H NMR chemical shifts of tribromomethane in several solvents, and its one-bond coupling constant, J_{CH}. Amine solvents: □; oxygen-donor solvents: ○; aromatic solvents: △; other solvents: ▽. (b) The relationship between ^{1}H NMR chemical shifts of trichloromethane in several solvents, and its one-bond coupling constant, J_{CH}. Curve (a): amine solvents; curves (b) and (c): see text; curve (d): aromatic solvent.

Further evidence for interaction involving the iodine atoms of HCl_3 was obtained from its infrared spectrum [158]. The stretching vibration of the C–H group was found to be unaffected by the formation of complexes with halide ions. This behaviour is distinctly different from that of the other trihalomethanes, for which ν_s(CH)—or ν_1 in Herzberg's notation—undergoes the frequency shift, broadening, and intensification expected on hydrogen bond formation.

irrational, unperfumed by *significance*, leading to nothing except the extinction of every virtue, and eventually mind itself.

This commitment of the few to tell the truth of war, despite the yearning of many for easier presentations, has been noted nowhere more bluntly or brilliantly than in Simone Weil's remarkable essay, *The Iliad, or The Poem of Force*. There, she writes:

> The true hero, the true subject, the center of the *Iliad*, is force. Force employed by men, force that enslaves man, force before which man's flesh shrinks away. . . . To define force—it is that *x* that turns anybody who is subjected to it into a *thing*. Exercised to the limit, it turns man into a thing in the most literal sense: it makes a corpse out of him. Somebody was here, and the next minute there is nobody here at all; this is the spectacle the *Iliad* never wearies of showing us. . . .

While the South found no epic poet to show us war as it is—the Civil War as it *was*, for the men of my family and the others who fought—I have found at least two memoirs of the truth that will haunt me all my days. One is the very quiet, coldly horrific battlefield photograph of Matthew Brady and Alexander Gardner. Along with many anonymous photographers who walked in the bloody paths cut by generals, these men produced an unforgettable record of the encounter between the camera, the nineteenth century's most important new technology of image-making, and industrialized war, the era's greatest invention for producing death. Dispersed in archives and libraries throughout the United States, the pictures still possess an undispelled gauntness, unsoftened by the billowing rhetoric of static, standardized mythology—the power to sweep away the wreaths of wishful thinking, and reveal the obliterating violence of war.

The other record lies in the writings of ordinary Confederate soldiers. It was while hunting for documents of the war by such men—witnesses who might not be as prone as their former generals to Lost Cause bombast—that I came across a long-buried story by my kinsman Samuel Elias Mays, lawyer in Pendleton and walker in deserted Charleston, sometime of the Hampton Legion, Confederate States Cavalry.

On the other hand, a vibrational mode of the CI_3 group—designated ν_2 by Herzberg—exhibits an increase in intensity of about two orders of magnitude. This is due to perturbation of the CI_3 group. The other trihalomethanes show a small frequency shift of ν_2 on complex formation involving the C–H group, but no appreciable change of intensity.

The fact that $HCBr_3$ has smaller complex shifts than $HCCl_3$ may be an indication that an unknown, but small, fraction of the tribromomethane molecule interact via the CBr_3 group. This postulate would certainly account for the smaller complex shifts; however, there is no evidence in the infrared spectrum for such a phenomenon, so the question remains open.

It is found that the one-bond (C–H) coupling constant in tribromomethane is linearly related to the induced shift in a number of solvents, although there are separate lines for aromatic solvents, amines, and all other solvents, all three of which intersect near the point for cyclohexane solvent [391] (see Fig. 3.4). The slope of the line for the aromatic solvents is evidently due to the magnetic anisotropy contribution of the solvent molecules to the shift of $HCBr_3$. The induced shifts in the amine solvents are greater than those in other donor solvents; this behaviour is reminiscent of that of $HCCl_3$ in the vibrational spectrum (see Section 3.1.5) in which only nitrogen donor species (and sulphoxide cause large changes in the stretching force constant of the C–H bond. A similar division of solvents is suggested by the present results and by the behaviour of the proton chemical shifts and one-bond coupling constants for trichloromethane

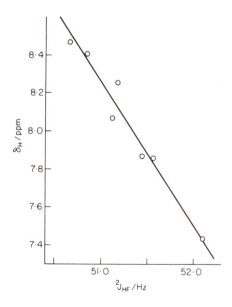

Fig. 3.5 The relationship between the ^1H NMR chemical shift of HClFClBr in several solvents, and its two-bond coupling constant, J_{HF}.

(also shown in Fig. 3.4). However, in the case of trichloromethane, the solvents of the classes ether, alcohol, ketone, amide and sulphoxide follow their own linear correlation (curve (b)) which is separate from that for chlorinated hydrocarbons, acid chloride, nitroalkane and nitrile (curve (c)).

The two bond H–F coupling constant in HCBrClF was found by Evans to exhibit a solvent dependence which is linearly related to the proton chemical shift variation in the same solvents [116] (see Fig. 3.5). However, this variation is not a large one and is not yet accounted for satisfactorily.

3.2.8 Miscellaneous

3.2.8.1 Dipole moments

Other than trichloromethane, very few dipole moments of trihalomethane complexes have been measured. Earp and Glasstone [113] obtained the following values: $HCBr_3/(C_2H_5)_2O - \mu = 6 \cdot 13 \times 10^{-30}$ C m; $HCBr_3/[(CH_3)_2O - \mu = 8.60 \times 10^{-30}$ C m. These values are smaller than the corresponding ones for trichloromethane by slightly more than the dipole moment difference of the free molecules. In common with trichloromethane, the results do not appear to shed much light on the nature of such complexes.

3.2.8.2 Molar Kerr constants

Le Fevre, Ritchie and Stiles have reported that the molar Kerr constants of trichloromethane and trifluoromethane at infinite dilution in benzene are much more negative than the values for the vapour phase [239]. They point out that such an observation is consistent with a species in which the dipole moment lies perpendicular to the directions of maximum polarisability. Thus, the preferred orientation of HCX_3 is with the C–H bond perpendicular to the plane of the ring of a solvent molecule.

3.2.8.3 Dielectric polarisation

Glasstone, in his measurements of the polarisation of trihalomethane mixtures with donor molecules, arrived at the following ordering based on the extent of complex formation [145]:

$$HCCl_3 > HCBr_3 > HCl_3 \gg H_3CCl_3$$
$$HCCl_3 > CBr_4 > CCl_4$$

The criterion chosen was the magnitude of the deviation of the apparent polarisation of one of the components as a function of $[(\epsilon_r - 1)/(\epsilon_r + 2)]^2$, from the corresponding value in a mixture with the non-polar cyclohexane.

Noting, years after the fact, what he recalled of his battles near Richmond, the elderly veteran makes no attempt to infuse his narrative with meaning, but instead allows the fragments of recollected experience to stand raw, disjunct, unexplained. He remembers mud and dust, and night, and the dying at night. Of his struggle across a sunless battlefield, he writes:

The cries of the wounded Yankees sound in my ears yet.

Many of them were, I have no doubt, drowned, as the ground and ditches were full of water, and some of them seemed to be strangling. . . . I was on the field the next day and saw the Yankee dead, which were strewed, as it seemed to me, all over the face of the earth and in all manners of positions and attitudes, and all dead.

Then the rain that had drowned the Yankees stopped; and was followed by drought. Within only a couple of days, it was exceedingly dry . . . and the dust was suffocating, rising in such volume as to almost hide the head of your horse from view . . . We could neither see nor breathe for the dust. The road was filled with wagons and artillery, which only served to make matters worse. Just before the top of the hill, a long gentle rise, we struck the battlefield and the first dead were our own men. They were scattered all through the bushes and would average about two dead men to every square rod for several hundred yards and all dead, the wounded having been removed.

But the worst sight was around a Yankee battery near a clump of pine trees. It was the Yankee supports of the batter and covered a space of about one acre. Here the Yankee dead literally covered the ground and in some places were piled on top of each other as many as three deep. I remember one dead Yankee particularly. He had been peeping around a tree and a bullet took him in the middle of the forehead exactly. Another man I saw and recognized, a dead Confederate. He was sitting up on the side of the road binding up a wounded leg with his breeches rolled up above his knee. He looked so natural that I stopped to speak to him, and a further look showed that he was dead. . . .

The formation constants which were deduced from these measurements are given in the appropriate tables.

3.3 Other H–C$\lessdot$ Compounds

3.3.1 Compounds with Halogen Substituents

It is usually found that C–H-containing compounds with substituents other than halogen form complexes with donor molecules, if at all, less readily than do the trihalomethanes. For example, it has been found that the formation constant for the pentachloroethane/diethyl ether complex is smaller (0·35–0·45; 0·36 mol mol^{-1}) than the value for the corresponding trichloromethane complex (0·40–0·66; 0·44 mol mol^{-1}) [113, 145]. This has been taken to mean that the C–H bond of the acceptor must be 'activated' in some way in order that hydrogen bond formation may occur.

In addition, the frequency shift of the stretching frequency, ν_s, of a number of heterogeneously substituted C–H compounds in mixtures with dimethyl sulphoxide and azine have been reported [7]. These frequency shifts are given in Table 3.22 (see p. 98). Unfortunately, the reference frequency in each case is measured in tetrachloromethane solution rather than in the vapour phase. Nonetheless, these shifts are seen in all the reported cases to be rather large, indicating substantial perturbation of the C–H groups. Indeed, several such frequency shifts are considerably larger than those of the trihalomethane complexes.

A rather more thorough study has been reported by Alley and Scott [8, 9] of complex formation constants for 1-hydroperfluoroheptane, $HCF_2(C_6F_{13})$. The values were measured at several temperatures in binary mixtures with donor molecules. The values obtained are given in Table 3.23.

3.3.1.1 Enthalpy of mixing

A number of values of $\Delta_m H$ for HCXYZ compounds have been measured [259], and are given in Table 3.24 (see p. 99). The relatively large values for 1,1,2,2-tetra-chloroethane may indicate the possibility of chelation: i.e. interaction of both C–H groups with an electron donor simultaneously.

3.3.1.2 C–H stretching mode

The frequency of this mode is found to decrease upon complex formation, as expected by analogy with the trihalomethanes. For pentachloroethane, ν_s is found at 2965 cm^{-1} in CCl_4 solution, but at 2895 cm^{-1} upon addition of tributylphosphine oxide [163]. 1,1,2,2-tetrachloroethane shows a smaller difference–2960 and 2925 cm^{-1}. The addition of the donor molecule in each case results in an increase in the intensity of ν_s, although less than is the case for trichloromethane.

In Samuel Elias's narrative, we are shown pure force, without the decorative trappings of exculpatory myth. The nostalgic, falsifying ideology would come later; seeding and reseeding itself in Southern imagination, mutating over time into deeds and words—flowering grimly as the night ride and white terror in the wake of war, then as a gathering of malice against New South conciliators around the turn of the century, later as the poisonous racism that still moves under the skin of American culture, erupting, from time to time, with the old ideological virulence, the same violent language. But in the memoir of Samuel Elias Mays, we are very close to the dumbstruck recognition of what war is to those who fight: a hollow moment in which *somebody was here,* as Simone Weil writes; and the next minute there is *nobody here at all.*

Out of this nothingness stumbled a generation of Southern boys who had known little of life save rumors of war, then the chivalric fever that swept the South at the war's outset, and the stark anxiety and mind-destroying violence of the battlefields.

The most fortunate were those who went home to the old order, however shaken by war and disrupted by Emancipation, however constricted by the early, radical onslaught of Reconstruction. Stepping from the crowded troop train at Mayesville, the town's weary young soldiers discovered the key guideposts of Southern existence still standing, and thus knew where and who they were. The Brick Church's massive pillars and tall walls, symbols of ancient stability rising over the cotton fields, stood intact; and the Squire waited on the steps of his manor house for his children to return. He still reigned over the intricate network of kinships and partnerships established before the war. And if many of the old cultural threads linking families and enterprises were frayed—the agrarian ideals of small-town living, for instance, and the conventional structures of church and law—they remained unbroken. Mayesville was destitute, but not prostrate—and, still less, devoid of the consolations of home-place, sustaining institutions, and loyalties.

Reliving Mayesville's postwar years through Sallie Burgess's narration of the Squire's life and offspring, and in more recent summaries such as those of Ethel Turner, the reader recognizes this intricate tracery of the rural Christian society, lending its gracious cohesion to the townsfolk as peace returned. Its gentle powers of involvement, we learn, were not lost even on those who came later to Mayesville, merely to make their fortunes in the reviving, commercially expanding community. The hardworking young Baptist newcomers, for instance, arriving in Mayesville by the early 1870s, did well. And as they settled and prospered, these people—as newly enriched Southerners did then and do now, for some reason—fell under the sway of vivid religious rigorism, of a kind foreign to the tolerant grain of life among the settled old Calvinist families. The Baptist ministers took to delivering regular tirades about "back sliding, profanity, intoxication, cockfighting, and dancing"—which in turn inspired many a young Baptist to defect to the Presbyterians. As denominational changes sometimes are, these conversions were partly principled, and largely opportunistic: for authority in Mayesville resided with the old Presbyterian elite, and everyone knew it. But, whatever their motives may or may not have been—I leave that for the Lord to decide—one fact is unshakable. The converts were abandoning the heresy of Fundamentalism, narrow and narrowing, to which Southerners have always been peculiarly prone, in favor of Southern Christianity in its more durable and sustaining mode: Trinitarian in faith, Reformed in polity and emphasis, Classical in matters of civic and private morality, *patiently* evangelical, *modestly* pious. These new men and women, that is, were enjoying the first fruits of old Southern civilization, as created and sustained in Mayesville before the Civil War, and alive after it.

The notes of Ethel Turner are miniature watercolors of this revived, reinvigorated small-town culture. We drop in on the Book Club, a genteel society chaired by Jesse Chandler and Daisy Witherspoon Muldrow, both scions of the old planter and merchant caste. We stand at the back of the crowd during gala gatherings convened by the Women's Civic League to dedicate a new grove of silver maples, or new gazebo at the cemetery. We join the annual spring festival under the lovely skies of South Carolina, and witness the crowning of the May Queen—a fertility rite at bottom, of the sort farmers and land folk in the West have celebrated since our ancestors in antiquity began to plant, though tidied up in Mayesville by Presbyterian and Southern Victorian ideas of propriety.

And so the families there went on, long into the twentieth century—farming, marrying their own cousins and the daughters of

3.3.1.3 Effects on the NMR spectrum

A number of solvent shifts of HCXYZ type compounds are given in Table 3.25 (see p. 100), as well as some complex shifts of dichloroethanenitrile. The latter confirm the usual observation that the interaction of a C—H group with aromatic donors becomes stronger with the substitution of alkyl groups, as indicated by the magnitudes of the complex shifts.

The solvent shifts in dimethyl sulphoxide become larger for ethanes as they become more highly substituted with electronegative groups such as halogen. Even the 1,1,2,2-tetrachloroethane solvent shift is smaller than that of penta-chloroethane, suggesting that the possibility of chelate formation involving the C—H groups of the former molecule may not be an important factor in its complex formation.

3.3.1.4 Freezing point diagrams

The freezing point diagrams of $HCCl_2CHCl_2$ (I) and $HCBr_2CHBr_2$ (II) with some aromatic compounds have been measured [252]. Formation of 1:1 compounds is indicated for (I) with 1,3-dimethylbenzene and probably with benzene and 1,4-dimethylbenzene but not with 1,2-dimethylbenzene; (II) forms a 1:1 compound with 1,3-dimethylbenzene but not with 1,2-dimethylbenzene.

3.3.2 Compounds without Halogen Substituents

Lemieux, Hayami and Martin have investigated [258] the complex formation properties of a series of acetylated glucopyranoside molecules with the anions of tetraethylammonium halide salts in ethanenitrile solution. They find that the ^{1}H NMR signals of H_1, H_3 and H_5 undergo a shift to lower field in the presence of halide ions, whereas the other signals are relatively insensitive to the presence of these ions.

They postulate the formation of a complex between a halide ion and the H_1–H_3–H_5 *region* of the carbohydrate molecule. From their data, the authors calculated the formation constants for such complexes, and the associated complex shifts for the three protons involved. For the pyranoside molecules with 4-substituted phenoxide groups as the β-substituent on the ring, the complex shifts of the protons on the phenyl ring were also determined wherever possible. All the results are given in Table 3.26.

3.4 Disubstituted Methanes, H_2CXY

It has long been felt that if three 'activating groups' such as halogen were able to 'activate' the C—H bond so as to cause it to participate in hydrogen bonding, then the presence of two such groups might well be sufficient to promote hydrogen bonding. Accordingly, considerable effort has been put into investigating the properties of complexes of dichloromethane and similar compounds.

3.4.1 Macroscopic Properties

3.4.1.1 Solubilities

Copley, Zellhoefer and Marvel [88, 89, 90] have reported the solubilities of dichloromethane in a wide variety of solvents. Their results are given in Table 3.27 (see p. 102). As these workers point out [88], the solubility of dichloromethane is very high in unassociated donor solvents (such as ethers, esters, amines and N,N-disubstituted amides) compared with the value of 0.311 mol mol^{-1} predicted by Raoult's Law. In highly associated solvents (such as alcohols and amides), the solubility is very low, while in weakly associated solvents (such as carboxylic acids, N-substituted amides and oximes) its solubility is intermediate.

The high solubility in unassociated solvents is ascribed to the effects of complex formation involving a $C-H\cdots X$ bond, where X is oxygen or nitrogen. The low solubility in associated solvents is attributed to the energy of the $X-H\cdots O$ bonds in the pure solvent being greater than that of the $C-H\cdots O$ bond which must replace it if dichloromethane is to dissolve.

The same workers found [89] that the solubility of dichlorodifluoromethane tends to be lower, in unassociated solvents, than that of dichloromethane. They accounted for this observation in terms of the inability of F_2CCl_2 to form hydrogen bonds. It should be noted, however, that the replacement of H by F results in the modification of properties other than the ability to form hydrogen bonds.

In a further paper [90], the same workers found a systematic variation in the solubility of H_2CCl_2 in a series of α,ω-dinitrile compounds, as given at the foot of Table 3.27. The short-chain dinitriles afford relatively poor solubility, which Copley *et al.* attribute to self-association of the solvent; the methylene hydrogen atoms adjacent to one nitrile group are further 'activated' by a second nearby nitrile group.

This self-association then ceases to be important when the chain length becomes great.

This postulate of solvent self-association is consistent with observation but poses more questions than it answers. The nitrile group is known to be weakly donating and yet this postulate involves considerable solvent self-association. The 'activation' of the methylene hydrogen atoms by a nitrile group four carbon atoms away seems most unlikely.

3.4.1.2 Viscosities of binary mixtures

Nigam and Mahl [286] point out that two relationships of viscosity to composition may be used as criteria of specific interactions in binary mixtures. First is

other important families in the vicinity, observing the basic rites of civilized life, consolidating the all-important cotton fields in the hands of men and women who understood the land, and its powers to nurture a steady sense of place. The essential elements of this agrarian way continued to hold, in fact, almost to the day Joe Chandler, Rutledge Dingle, and I arrived in Mayesville—almost. Not long before, the last Mayes descendant to dwell on the site the old Squire cleared for his house had sold off what remained of the family's land, and died shortly thereafter. The symbolic figure of a Mayes man standing at the head of clan and kinship—the role created by the old Squire 170 years before—was gone; and the last public symbol of coherence that had given a nourishing sense of order to the town, and an idea of place and home to the returning veterans of the Civil War, had vanished forever.

Unlike the young soldiers of Mayesville, John Matthew Mays came home to almost nothing, other than wreckage and urgent need. What farming he did on his mother's Saluda River lands in the summer of 1865 was undertaken to feed her and his unmarried younger sister, but with no intention of restoring the holdings or traditional life of his family by growing food or cotton all his days. Many of John's cousins and returning neighbors did do just that, adjusting to the new circumstances and refounding agricultural and mercantile fortunes, and good lives, over the next several years—and were being urged to do so by important voices in John Mays's neighborhood. The vigorous local secessionist Wyatt Aitken, for example, turned into a New South activist upon hearing of the Confederacy's defeat, and soon thereafter founded a monthly magazine called the *Rural Carolinian* to trumpet his views. "The times have changed," wrote Aitken, "and with them must change many of the habits of our people. . . . Forget the past; anticipate the future! . . . Live and farm as if slavery never existed . . . Cease your infatuation with cotton!"

It was a message broadcast by Aitken and men of like mind throughout the postwar South, and heeded by some farmers in search of continuity with the past best guaranteed by such changes— ones that would not really touch agrarian culture's heart or stop its pulse in Southern civilization. In light of my great-grandfather's later success at trade and banking in east Texas, I am sure he could have

the excess viscosity, η_E, which is defined as

$$\eta_E = \eta_{obs} - (x_1 \eta_1 + x_2 \eta_2) \tag{3.20}$$

It is expected that η_E will be positive when strong interactions result in complex formation. In the absence of such interactions, η_E is usually negative.

The second criterion is based on an interchange parameter, d, which is defined by the equation

$$\ln \eta_{obs} = x_1 \ln \eta_1 + x_2 \ln \eta_2 + x_1 x_2 d \tag{3.21}$$

and is regarded as a measure of the strength of interaction between the components.

To confirm these expectations, the values of η_E and d were determined at a series of concentrations for the mixtures benzene/pentane, benzene/hexane and cyclohexane/pentane. In these cases, both parameters have large negative values, indicative of a lack of strong interactions. On the other hand, mixtures of trichloromethane with propanone, diethyl ether and 1,4-dioxane were found to exhibit large, positive values of η_E and d, confirming the presence of strong interactions resulting in complex formation.

Mixtures of dichloromethane with benzene; methylbenzene; 1,2-dimethyl-benzene; 1,3-dimethylbenzene and 1,4-dimethylbenzene exhibit intermediate behaviour – small negative values in benzene mixtures becoming small positive values in dimethylbenzene mixtures. Hence, intermolecular interactions of intermediate strength are indicated, increasing in the order: benzene $<$ methylbenzene $<$ dimethylbenzenes.

3.4.1.3 Enthalpies of mixing

These have been measured for several H_2CX_2 compounds with donor molecules, and are given in Table 3.27. Evidently, the amide molecule is a more effective donor than the ether, and the strength of the interaction seems to become smaller as the electron withdrawing ability of the substituent X increases.

3.4.1.4 Formation constants

The complex formation constants of H_2CXY, where X, Y = Cl, Br, with several donor molecules are given in Table 3.28 (see p. 102). These formation constants are all fairly small and exhibit the expected trends except that dioctyl sulphide forms complexes with these acceptors more extensively than does dioctyl ether. This seems anomalous, since the oxygen atom is generally recognised as a much better donor atom than sulphur. Perhaps the greater size of a sulphur atom enables it to interact more effectively with both C–H groups of the acceptor molecule simultaneously. Such a consideration might be expected to lead to a more negative enthalpy of formation of the sulphide complexes; as is seen in the next section, this expectation is not borne out by observation.

made such switches and alterations, had he wanted, and used his fruitful land in much the way his Uncle Samuel had done years before, as a springboard into South Carolina politics, law, or commerce. But John was born too late to share Old South pieties, and too ambitious to spend his young life trying to resurrect such ideals; his experience of war only snapped conventional bonds to the soil that were already tenuous. As soon as he had ensured that the bodies and souls of his sister, mother, and her husband would stay together, he made ready to leave South Carolina, and walk a path different from that of his ancestors and some contemporaries—free of old allegiances to Southern existence, without dreams of recapturing and rebuilding what had been lost in virgin territory. He did not seek advice in visions of past or future. Unlike the divine guidance that went before Israel in its long journey out of Egypt into a promised land, the pillars of cloud and fire lay *behind* John Mays—so much acrid smoke rising over the burned cities and fields and ruined ambitions of the Old South, to which he would never return.

By the end of 1865, not yet twenty years old, John had married Alice Starnes, daughter of a prominent plantation family in Laurens County, across the Saluda. And in the autumn of the same year, he loaded his new wife and his kin into a wagon and set out west, toward the border counties of east Texas, just over the line from northwestern Louisiana. How he got there—whether by packet from Georgia to the port of New Orleans, then by paddleboat steamer up the shallow Red River to Shreveport, or overland to the same destination—I do not know. But whatever his route, John and his family traveled it with throngs of other veterans of the Civil War, similarly lured from the devastated older South by the promise of new farmland and opportunity in the broadly rolling, beautiful Texas hills lying westward from Shreveport. Carrie, the first child of John and Alice, entered the world in November 1866, near the tiny farming and ranching community of DeBerry—a Texan, and the first of my close Mays relations to be born outside the South since the sixteenth century.

Despite its adjacency to the old Southern state of Louisiana, east Texas *is* different. The years of Texas independence were not many, but the culture of the republic left an indelible tattoo on Texas souls and land. Anyone who spends much time there knows this is not the

3.4.1.5 Enthalpies of complex formation

The values of $\Delta_\phi H$, and other thermodynamic parameters, for H_2CXY complexes are given in Table 3.30 (see p. 103). In all cases, $\Delta_\phi H$ is found to be small and negative. Unfortunately, no vapour phase values are available, with the result that the relationship of ΔH_1 (measured in solution) to ΔH_4 (in the absence of solvation) is not known (cf. Section 3.1.4).

3.4.2 Effects on the Vibrational Spectrum

The vibrational modes of the CH_2 moiety are necessarily more complex than those of CH. In particular, the stretching modes exhibit two characteristic absorptions corresponding to symmetric and antisymmetric motion, rather than a single one.

Several groups of workers have studied the effects of the presence of electron donor molecules on these stretching modes. The intention of discovering large shifts to lower frequency, usually a diagnostic feature of hydrogen bond formation, has been largely unfulfilled. While these vibrations are found at slightly lower frequency with donor molecules, the magnitudes of the frequency shifts are not large enough to be unambiguous about the nature of the perturbation [7, 118, 208, 287]. Similarly, the frequencies in the Raman spectrum are affected very little at very low temperatures in crystalline H_2CCl_2 [39].

3.4.2.1 Dihalomethanes

Kagarise [204] observed that the symmetric stretch of H_2CCl_2, ν_1, changes very little in frequency, half-width, or intensity from tetrachloromethane solvent to a 1:1 propanone/CCl_4 solvent. On the other hand, the antisymmetric stretch, ν_6, changes frequency by $+9$ cm^{-1}, and increases markedly in intensity. The author points out that these results indicate that the dipole moment change perpendicular to the molecular axis of symmetry is strongly affected by the (presumed) formation of a complex with propanone, while the change parallel to this axis is unaffected.

Assuming 1:1 complex formation between H_2CCl_2 and propanone, a formation constant of 0·05 mol dm^{-3} is obtained. The intensity difference in ν_6 between 'complexed' and 'free' dichloromethane is a factor of 14·5 (cf. $HCCl_3$/propanone: intensity change = 20·5).

Kanbayashi [208] has observed an analogous increase in the intensity of ν_6 for dichloromethane on self-association in CCl_4 and CS_2 solutions. A similar behaviour, accompanied by decreasing intensity of ν_1, has been observed in the infrared spectrum by Josien, Fuson and Lafaix [196]. No such change was observed for dimethoxymethane-d$_6$, $H_2C(OCD_3)_2$. The Raman intensity of ν_6 is not concentration dependent.

Evans and Lo [118] find the same lack of concentration dependence for the

South, and east Texans know they are not Southerners. At least in my growing-up days, they seemed to be men and women less tragically resigned than Southerners to the inevitable fractures in time and fortune, more enterprising, largely free of durable allegiances to agrarian ideals of life and manner. As a child, when visiting Greenwood, I would occasionally go with Sister Erin or Aunt Vandalia or Uncle Alvin to the railway town of Waskom, only five miles away, just beyond the Texas line. I remember it as a special adventure, because the people were different—taller and stronger than Southerners, it seemed to me, like Tom Mix or Gene Autry; and were often seen in cowboy boots, which nobody in Greenwood wore. The houses were smaller, more scattered and weather-beaten, the men quieter and somehow more relaxed than the Southerners in Greenwood. In Waskom, ladies rode around in and even drove pickup trucks—something no Southern woman, old or young, in Greenwood's stable matriarchy would do, least of all Sister Erin or Aunt Vandalia.

To a certain extent, these impressions of extraordinary difference were the fantasies of a child who had seen and loved many Western movies, and thought of *Texas* less as a place than as a permanent rodeo, or a shoot-out at high noon. But what I felt was not all illusion. The region of east Texas in which Carrie Mays was born, and, in 1870, her brother Robert, had already been gradually peopled in the antebellum era by relative newcomers to America, frontier-minded folk whose formative experiences in the eighteenth century had been the hardscrabble upcountry, the hardtack little farm in a deep Tennessee valley, and subtle disdain for lowland planters. Into this scatter of old east Texans—Scots-Irish families from the Southern mountains, Missouri farmers, graziers, and ranchers from upcountry Virginia—the Civil War propelled a greater number of former Confederates and their children, many of whom had left behind memories of the Old South on their barren plantations. Nothing suggests that John Matthew Mays or his stepfather, William Moultrie Griffin, for example, had any intention of settling down quietly to farm in east Texas according to old Southern ways. Both did what their descendants, by and large, have always done when finding ourselves in a difficult spot, without money or contacts with some settled tribe. They sat down, decided what the other newcomers needed and wanted, then figured out how to deliver it to them.

Law was needed among the immigrants of DeBerry, so John taught himself the basics of law and enforcement, and became a justice of the peace. Schooling was wanted by the settlers, so he became a teacher. The proclamation of Methodist doctrine was also wanted—a service John left to William Moultrie Griffin, who, in my family's stories, was an eloquent itinerant preacher to the scattered Christian congregations that met for prayer in tents or makeshift chapels alongside the narrow dirt roads of east Texas. But even more urgently than sacraments or schools, the new east Texans needed manufactured goods—tools and medicines, cloth and sugar and crockery, and other essentials that could not be made on a homestead. These requirements focused John Mays's ambitions, and those of Moultrie Griffin, who had known storekeeping from the war years, when he ran a Confederate Army commissary in the South Carolina town of Ninety Six. By the beginning of the 1870s, John was driving mule-drawn wagons laden with supplies up from Shreveport, through Greenwood, and into Texas, where Moultrie was selling them. The flowering enterprise was further enriched when John's sister Anna Lucretia married a merchant already established at Bethany, a few miles east of DeBerry on the Texas border with Louisiana.

It is possible that my great-grandfather might have stayed in this expanding circle of successful merchant-kinsmen had not personal disaster struck deep into the heart of his own family in 1871.

John had come with his wife, Alice, from postwar hardship in South Carolina into the more primitive conditions of east Texas, where the old pattern of American settlement was reproducing itself in the clearing of homesteads near creeks for water to drink and near shallow navigable streams. They had just begun to prosper in that new terrain when yellow fever, the South's ancient enemy, rose again from the murk and silence of marshes and flooded creek bottoms among the hills to gather in its occasional harvest of souls.

In most seasons, the appearance of the mosquito-borne plague was brief, and it was selective and unpredictable. After carrying away whole villages, or making a few people gravely but briefly ill, the menace would then recede as quickly as it had come, and would remain in abeyance for a couple of seasons. It might return with devastating force. Nothing east Texans had ever seen, however, matched

two stretching modes of H_2CCl_2, H_2CClBr and H_2CBr_2 in the Raman spectrum. On the other hand, the infrared intensity of ν_6 increases very much in ethanenitrile, propanone and dimethyl sulphoxide solvents. The intensity of ν_1 increases in dimethyl sulphoxide solvent, but not in propanone; it decreased in ethanenitrile solvent.

From a normal co-ordinate analysis, Evans and Lo conclude that, in the vapour phase, $\delta\mu_{CH}/\delta r_{CH}$ is essentially zero in H_2CCl_2, and that $\delta\mu_{CCl}/\delta r_{CH}$ is large and accounts for the observed intensity of ν_1. In dilute solution in tetrachloromethane, they point out that an increase in the value of $\delta\mu_{CH}/\delta r_{CH}$ accounts for both the increase in ν_6 intensity and the decrease in ν_1 intensity. They regard the interactions between dichloromethane and other molecules in the condensed phase as partaking of a charge-transfer character: viz. $\bar{C}\,H\cdots\overset{+}{X}$. The more strongly interacting solvents, therefore, produce a greater enhancement of ν_6. As $\delta\mu_{CH}/\delta r_{CH}$ increases in magnitude, it first leads to a reduced intensity of ν_1 by cancellation of the cross-term $\delta\mu_{CCl}/\delta r_{CH}$, as in H_2CCl_2 and CH_3CN solutions. In the more polar solvents, the first derivative becomes predominant and the intensity increases, as in dimethyl sulphoxide.

Similar intensity behaviour was noted in the cases of chlorobromomethane and dibromomethane, although the vapour-phase intensity of ν_6 is increasing in the progression $H_2CCl_2 < H_2CClBr < H_2CBr_2$. The derivative $\delta\mu_{CH}/\delta r_{CH}$ appears to increase in this series, while $\delta\mu_{CBr}/\delta r_{CH}$ is likely smaller than the cross term involving the CCl bond. Indeed, in diiodomethane, the intensity of ν_6 exceeds that of ν_1.

In contrast to the consistent observation of small frequency changes in ν_1 and ν_6 under different conditions, Oi and Coetzee [290] have found, for H_2CCl_2, large solvent dependences of the C—Cl stretching frequencies ν_3 and ν_9 (symmetric and asymmetric, respectively). The shifts from the vapour-phase wavenumbers, $724\cdot5$ cm^{-1} and 759 cm^{-1}, are given in Table 3.31 (see p. 104). These workers reported that, although $\Delta\nu/\nu$ for C—Cl stretching modes in molecules such as 1-chloropropane and 2-methyl-1-chloropropane exhibit a linear dependence on $(\epsilon_r - 1)/(2\epsilon_r + 1)$, where ϵ_r is the solvent permittivity relative to that of a vacuum, the values of $\Delta\nu/\nu$ for H_2CCl_2 do not show such a dependence. In the electron-donor solvents 1,4-dioxane, diethyl ether, ethyl ethanoate, propanone, ethanenitrile, nitromethane, and triethylamine, the frequencies do not depend in this simple manner on solvent permittivity, as may be seen in Figure 5 (for $HCCl_3$) and Figure 2 (for H_2CCl_2) of reference [290]. This was accounted for on the basis of the formation of hydrogen bonded complexes with the solvent molecules, similar behaviour being observed in solutions of trichloromethane.

3.4.2.2 Effects on the donor molecule

The C=O stretching mode of ethyl ethanamide is found to be at lower wavenumber in dichloromethane solution (1671 cm^{-1}) than in tetrachloromethane (1687 cm^{-1}) [218]. This effect is analogous to that in $HCCl_3$ and $HCBr_3$

solvents and presumably has its origin in perturbation of the donor molecule accompanying complex formation.

Klemperer, Cronyn, Maki and Pimentel [218] have measured the peak positions and intensities of the N–H stretching mode in N-ethyl ethanamide in several solvents. The intensity ratio of monomer to polymer was found to be higher in the solvent H_2CCl_2 (as also in $HCCl_3$ and $HCBr_3$) than in the weakly interacting solvents CCl_4 and CS_2, or in 1,1,1-trichloroethane ('methylchloroform'). This change in intensities is due to a lessened extent of self-association in H_2CCl_2 because of solute–solvent interactions. Since the C–Cl···H–N interaction accounts for little of this effect (cf. H_3CCCl_3), the C–H···O=C is taken to be primarily responsible for the observations.

A somewhat contrasting result was obtained by Bellamy *et al.* [30] who measured the X=O stretching frequency of a number of compounds containing S=O, P=O, O–N=O, and N–N=O groups. These vibrational modes were found to be solvent dependent, where the solvents included the dihalomethanes as well as mono- and tri-halomethanes among others. No unusual effects were observed to necessitate the postulate of hydrogen bond formation in solvents such as the di- or trihalomethanes. On the other hand, Philippe and Cléchet [306] attribute the S=O stretching bands of dimethyl sulphoxide in dichloro- and dibromomethane to hydrogen-bonded complexes of 1:1 (1075 and 1072 cm^{-1}) and 2:1 (1051 and 1055 cm^{-1}) stoichiometry.

3.4.3 Effects on the NMR Spectrum

The solvent shifts of a number of H_2CXY compounds have been measured, and are given in Table 3.32 (see p. 104). In all cases, these shifts are smaller than the corresponding ones for trisubstituted methanes. The disubstituted methanes bearing chlorine and/or bromine exhibit somewhat larger solvent shifts than do those with iodine or CN.

In the case of 1-substituted ethanes, the solvent shifts in dimethyl sulphoxide are found to increase with increasing number of electron withdrawing substituents on the adjacent carbon atom, this observation being consistent with the greater extent of complex formation found.

The complex shifts of dichloromethane with four aromatic donors and 1,4-dioxane are also listed in Table 3.32. The complex shift with 1,4-dioxane is smaller than that of trichloromethane (0·47 ppm *vs.* 0·65 ppm). Possibly the extent of shielding by the magnetic directing of the electric field of the donor atom along the molecular axis of H_2CCl_2 rather than along one C–H bond results in this smaller shift, since field-induced shifts are dependent upon the angle between the field direction and the C–H bond [55].

The complex shifts with aromatic donors are about the same as, or slightly smaller than, those of trichloromethane. The extent of shielding by the magnetic anisotropy of the aromatic molecule is evidently similar for the two acceptor molecules.

The chemical shifts of a number of propargyl compounds (i.e. 3-substituted propynes-1) have been measured [82, 171, 399]. While the acetylenic proton exhibits substantial chemical shift differences in different solvents (see Chapter 5), the methylene proton (3-proton) shifts are also found to be solvent-dependent.

The solvent shifts of the 3-proton of 3-chloropropyne-1 and 3-bromopropyne-1 are given in Table 3.33 (see p. 107). Both the papers from which these data are taken conclude that hydrogen bonding involving the alkyne C—H group is responsible for the interaction leading to complex formation. However, as Coleman *et al.* have pointed out [82], the methylene shifts in these compounds change by almost twice as much as the alkyne proton shifts on complex formation with benzene. It seems reasonable to conclude that interaction with solvent may occur at the methylene hydrogen atoms as well as at the alkyne one. The model compound 1,4-dichlorobutyne-2, which has no alkyne hydrogen, is found to interact with benzene to the same extent as does 3-chloropropyne-1, showing that this interaction is not dependent on the presence of the alkyne hydrogen.

The following formation constants of complexes with benzene were found at 303 K:

$$HC{\equiv}CCH_2Cl \qquad K(1:1) = 0 \cdot 924 \pm 0 \cdot 005 \text{ mol mol}^{-1}$$

$$HC{\equiv}CCH_2Br \qquad K(1:1) = 0 \cdot 869 \pm 0 \cdot 005 \text{ mol mol}^{-1}$$

$$ClCH_2C{\equiv}CCH_2Cl \qquad K(1:1) = 2 \cdot 56 \pm 0 \cdot 04 \text{ mol mol}^{-1}$$

$$K(1:2) = 1 \cdot 55 \pm 0 \cdot 10 \text{ mol mol}^{-1}$$

3.4.4 Other Disubstituted Methanes

Aksnes and Songstad [3] have measured the infrared stretching frequencies of CH_2 groups attached to a carbonyl group and an onium group. The shifts in these frequencies are taken to be due to hydrogen bond formation with the bromide counter-ion in each case; however, the frequency is found to increase or decrease, depending on the nature of the onium group and the other carbonyl substituent. It would appear that the observed frequency differences owe more to the changes in substituent than to possible hydrogen bonding.

3.4.5 Other Evidence

3.4.5.1 Crystal structures

The presence of close contact between a C—H bond and an electron donor atom in a molecular crystal is often taken as evidence of the existence of a hydrogen bond [368], although the validity of this criterion has been questioned (see Section 5.2.1).

Examples of such close contacts involving CH_2 groups are known. The hydrogen atoms of the methylene groups in the molecules are reported [62] to be in close contact with carbonyl oxygen atoms (227 and 226 pm). This is cited as evidence of intermolecular hydrogen bonding.

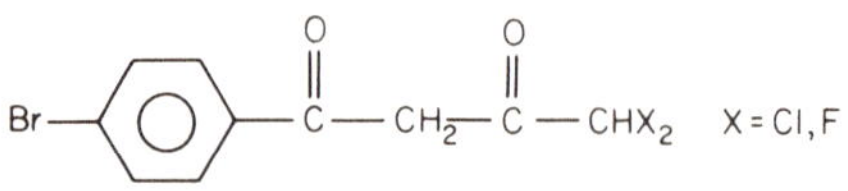

Another example of intermolecular close contact is reported [294] for the molecule shown here:

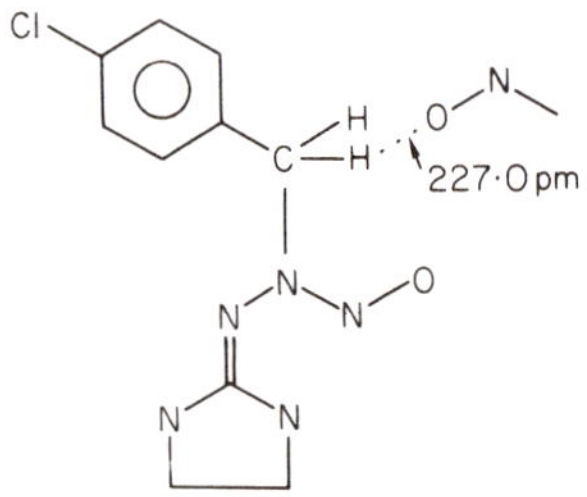

Manohar and Ramaseshan report a number of intramolecular close contacts between C–H groups and oxygen in the molecule of echitamine iodide [255]. These involve carbon atoms in different states of hybridisation—i.e. sp^3 and sp^2—and different degrees of substitution.

3.4.5.2 Polymer glass temperatures

Raitt has suggested that the CH_2 groups in chains of poly(chloroethene) undergo hydrogen bonding to the chlorine atoms of adjacent chains [317]. This proposal is based on the fact that this polymer has a glass temperature of 353 K whereas that of polypropene is only about 273 K.

3.4.5.3 Dielectric saturation

Chelkowski has found that the molecules 1,2-dichloroethane, 1,2-dichloropropane, 1,4-dichlorobutane and 1,4-dibromobutane exhibit differently shaped dielectric saturation curves in tetrachloromethane from those in benzene [73]. He speculates that the different shapes may be due to hydrogen bonding of the CH_2 groups with the π electrons of benzene.

3.4.5.4 Azeotropic mixtures

Ewell and Welch have investigated the phenomenon of maximum boiling (azeotropic) mixtures of chloroparaffins with donor liquids [119]. They find that compounds of the type H_2CCl_2 and CH_2ClCH_2Cl, among others, do

the vast outbreaks of the early 1870s; and nothing John Mays had encountered in his short life would mark him more strongly than the sudden passage, in a matter of hours, of both Alice and Carrie, aged five, from health into the hell of chills and blinding headaches that herald yellow fever. (The infant Robert was untouched.) The usual week of exhaustion and vomiting and the dreaded trademark jaundice followed the sudden onset of symptoms. After some ten days of grave illness, the final onslaught came, bringing delirium and frantic twitching that relented only in the last hours of life left to Alice and little Carrie Mays.

John laid his wife and daughter in the Methodist burying ground near DeBerry, marking their graves with simple surrounds of stone. The site of the resting place—even the name of John's first daughter—were unknown to me until, late one dry spring afternoon, I found their graves in the quiet Texas hills. Long after their deaths, certainly in the twentieth century, someone had returned and put up an impressive monument to them. I do not know who did so, though the man likeliest to have remembered them, and wished to mark his remembrance of mother and sister in stone, is John's son Robert.

The raiser of the stone was almost certainly not John. From the day of their death until the day of his own, in 1923, he never spoke of Alice Starnes or their daughter. Mentioning the names appears to have been taboo among his other children, and even into the generation of my father and Aunt Vandalia—though my aunt did break this family code on occasion, telling me almost all I know of these deliberately forgotten relations. For in the cluster of family names attached to faces in old photographs and recollected tales of other relations in our postwar migration to Texas—John Matthew Mays's second wife, Georgia Trammell, John's mother and sister and many others—Alice Starnes's name is an ellipsis, a genealogical blank. Vandalia noted her birthplace as Lawrence, Georgia. There is no such place, though the record keepers in my family and the genealogists in my aunt's circle never bothered to check. The new family Uncle Robert grew up with—his stepmother Georgia and her nine children by John Mays—seems not to have counted him as one of themselves. He rapidly became remote, disappearing from the letters and photographic records that have survived. And though he became an attorney in east Texas and, later, a well-off hotelier and businessman in Shreveport, only a few miles from Greenwood, he was very rarely named among us. My great-grandfather was a man able to forget forcefully—he had certainly abandoned South Carolina with vigor and finality—and also one capable of enforcing forgetfulness on his family, even encouraging them to shun a son, our little-known Uncle Robert—who, a relative once told me, was, after all, never a *real* Mays.

By all accounts, John Matthew Mays—Captain Mays to his children, Captain Johnny to his male buddies—was a domestic tyrant, a superb businessman, a stiffly well-mannered gentleman in all matters that did not involve money, and a forceful New Southerner in all matters that did.

After the death of wife and daughter in 1871, the Captain welded himself into an old, elite mercantile family near the east Texas county town of Henderson, by marriage, in 1878, to my large and formidable great-grandmother, Georgia Monroe Trammell. Again like the Squire, he quickly created the raw material of dynasty by fathering nine children, who were to grow increasingly eligible for propitious marriages as John's social standing rose in Henderson and his wealth increased. His conspicuous mark on the Texas landscape was architectural. To this day, the most eminent commercial buildings on the unfortunately modernized central square of Henderson— its nineteenth-century storefronts concealed behind tiresome façades imposed some years ago, the once-prominent county courthouse gone, after destruction by fire and reconstruction on an ignominious side street—are tributes to John's success: the Mays & Harris Store, in John's day Henderson's smartest provisioner of drygoods to townsfolk made rich by cotton and oil, and his tall First National Bank, which he co-founded and served as president.

The sons of Captain Mays luxuriated in the *fin-de-siècle* culture their father had created for them in Henderson, indulged reluctantly by both parents—and taking perhaps too much ease in it. None shone during later business careers, other than the half-brother Robert. The daughters were different, made of sterner stuff—perhaps because, given the imperious father they were obliged to survive, they really had no other choice. Of these women, Great Aunt Helen,

Clear cut, Carolina low country

form such mixtures with donor liquids. This is taken as a criterion of the presence of 'active' hydrogen, capable of forming hydrogen bonds.

3.5 Monosubstituted Methanes, H_3CX

3.5.1 General Evidence

The reported incidence of hydrogen bonding involving the hydrogen atoms of methyl groups is very small. This has led to the supposition that 'activating' substituents must be present on a carbon atom before its C–H bond may participate in hydrogen bonding. However, there are several instances in which hydrogen bond formation by a methyl group has been proposed to account for experimental observations.

Pauling mentions, in his monograph on *The Nature of the Chemical Bond* [297] that self-association of pure materials such as water, ammonia, and hydrogen fluoride by means of hydrogen bonds leads to unexpectedly high values of their melting points and boiling points. He points out further that the boiling points of some methyl-containing compounds are substantially higher than those of their trifluoromethyl analogues. He correctly suggests that these differences can be explained by the assumption of hydrogen bond formation, although of course they do not prove it.

	Boiling Point/K
$F_3CC(O)Cl$	273
$H_3CC(O)Cl$	324
$(F_3CCO)_2O$	293
$(H_3CCO)_2O$	410

A review article by Dippy [107] suggests that a methyl group situated *ortho* to an electron-donor substituent on a benzene ring may participate in intramolecular hydrogen bonding. Thus, the high dissociation constant of 2-methylbenzoic acid relative to those of the 3- and 4-methyl acids can be attributed to formation of a hydrogen bond in the anion of the acid. Once again the explanation is consistent with the evidence but is not proven by it.

Dissociation constants of methylbenzoic acids, normalised to K for benzoic acid

	$K(CH_3C_6H_4COOH)/K(C_6H_5COOH)$
2-methyl	1·97
3-methyl	0·87
4-methyl	0·68

the last child born to the Captain, has always fascinated me most, while eluding my curiosity as deftly as she resisted the prying of everyone else in the family.

By 1922, when she was in her early twenties, Helen had left Texas forever, and enrolled at Columbia University in New York City. She spent the middle years of that decade traveling in Europe—among Aunt Vandalia's things I found picture postcards dispatched by Helen from Budapest and Constantinople—and ended up in the fledgling Soviet Union, where she worked a few years, it seems, for an agency tending children orphaned during the Revolution of 1917 and the subsequent civil war. She was certainly back in New York by 1929, taking more courses at Columbia. There was then a quiet family scandal in the early 1930s—recalled by Aunt Vandalia only by indirections and whispers—when Helen became the companion of a well-off Manhattan woman, with whom she lived, in a small, conventionally well-appointed apartment on New York's Upper East Side, until Helen's death in 1963.

Beyond these bare details, I know almost nothing about my aunt—less, in fact, than I know about my eighteenth-century ancestors, and even some seventeenth-century cousins—but from childhood onward I have admired her independence, and the spirit of adventure and curiosity that took her abroad in the world. My last attempt to discover more about her, in 1967—I was particularly interested to know more about Helen's adventures in the Soviet Union—came to nothing. For by the time I visited my late aunt's lifelong friend at their apartment in New York's East Seventies, the antics of certain cousins—who had been filing affidavits and such in the belief that they were entitled to the considerable wealth Helen had accumulated, apparently through shrewd investment—had made her elderly companion leery of young Southern men named Mays. Nothing beyond forgettable chitchat was exchanged during that visit in 1967, which was conducted under the watchful gaze of the lady's lawyer, who never spoke.

Yet of all the children of Captain Mays, Helen was perhaps the one most like him. She left Texas as resolutely as he had left the South, and abandoned small-town ways with the same crisp finality he had discarded rural ones. And by settling comfortably and openly with another woman in a long, affectionate partnership—rejecting

The liquid–vapour equilibrium data at 318·15 K of Brown and Smith [52] show that the excess Gibbs free energy of nitromethane/propanone mixtures has a negative value and shows a change in sign of the deviation of the activity from the Raoult's Law value at a mole fraction of propanone of 0·35. The second virial coefficient B_{12} was found to be $-3·83$ dm^3 mol^{-1}, a value which is more negative than that of either component alone. It was concluded that the indicated interaction might be accounted for by hydrogen bonding, but that this conclusion was only tentative.

$$O_2NCH_3 \cdots O=C(CH_3)_2$$

Kebarle and co-workers [212] have obtained important results from mass spectrometry concerning the interactions of molecules with ions in the vapour phase. In particular, the reaction,

$$[X(H_3CCN)_n]^- \rightleftharpoons [X(H_3CCN)_{n-1}]^- + H_3CCN \tag{3.22}$$

where X = F, Cl, Br, I, has been studied for values of n from 1 to 5. The enthalpy of formation of each complex, as well as the entropy and Gibbs free energy change, was obtained from the temperature dependence of the appropriate equilibrium constant. The enthalpies of formation of the monosolvated ion complexes are: 66·9, 56·1, 54·0 and 49·8 kJ mol^{-1}, respectively. Because these values are considerably smaller than those for solvation of the corresponding alkali metal ions by ethanenitrile, the authors conclude that ion dipole interactions probably predominate in the solvation process, no explicit consideration of hydrogen bonding being necessary.

An X-ray determination of the crystal structure of dimethyl ethanedioate has revealed that three intermolecular C$\cdots$O distances of 335, 354 and 357 pm correspond to coordination of carbonyl oxygen atoms around a methyl group approximately in the directions of the C–H bonds [109]. The authors, Dougill and Jeffrey, suggest a weak association between the carbonyl and methyl groups. They point out that the melting point of dimethyl ethanedioate is anomalously high among dialkyl ethanedioates, although the boiling point is not. Hence, the intermolecular interaction is present in the solid phase but unimportant in the liquid.

Ethanedioic acid esters	Melting point/K	Boiling point/K
dimethyl	327	436
methyl ethyl		447
diethyl	232	458
dipropyl	228·9	487

3.5.2 Spectral Evidence

The frequencies of the n → π* transition in the carbonyl compounds propanal, 2-methylpropanal, and butanal have been found to exhibit a shift to higher frequency in the polar solvents diethyl ether, methyl ethanoate and ethanenitrile, the magnitude of the shift increasing in that order [293]. The shift magnitude is linearly related to the factor $(\epsilon_r - 1)/(\epsilon_r + 2)$ and hence is presumably due to polar interactions of the solvent molecules with the solute molecule in its ground state and excited state. This behaviour contrasts with the shifts in hydrogen bonding solvents such as trichloromethane and ethanol where the frequency shifts are much larger than expected from considerations of polarity. The conclusion reached is that molecules such as ethanenitrile and methyl ethanoate do not form hydrogen bonds with the carbonyl group.

The spectrum of azine-1-oxides at 270 to 285 nm undergoes a shift to higher frequency whenever a methyl substituent is present in the 2-position [188]. Ikegawa and Sato suggest a structure of the type to account for this shift.

However, in a series of substituted azine 1-oxides, this shift does not appear to be consistent, particularly when further methyl groups are present in the 3, 4, or 5 positions, so this conclusion must be regarded for the moment as doubtful.

The infrared spectrum of ethanenitrile containing dissolved salts has been measured by Perelygin [299] and by Gadzhiev and Pominov [138]. The C–H stretching frequencies are modified because of solvation of the cations Li^+, Co^{2+}, Mg^{2+} and Na^+. In addition, whenever the anion is iodide, further bands appear, suggesting hydrogen bonding of the methyl group with I^-. These anion bands are absent from the spectra of perchlorate salts. The C–H stretching bands at 2944 and 3002 cm^{-1} for $NaClO_4$, are accompanied by bands at 2924 and 2982 cm^{-1} for NaI solution. The same bands for LiI appear at 2930 cm^{-1} and about 2990 cm^{-1} (shoulder).

In the Allerhand and Schleyer survey of C–H groups as proton donors in hydrogen bonding [7], they note that 1,1,1-trichloroethane and iodomethane give no indication of hydrogen bonding to the strong electron donors dimethyl sulphoxide and azine. Similarly, the spectrum of dimethyl sulphoxide, $(CH_3)_2SO$, is independent of concentration in tetrachloromethane solution, showing the absence of self-association. Nitromethane and ethanenitrile, on the other hand, do exhibit changes in the vibrational spectrum on adding an electron donor. Since the changes are not typical of hydrogen bonded systems, they are attributed to other types of association.

c

In his study of solvent effects on the carbon–hydrogen coupling constant of trichloromethane, Evans found [116] an increase in $^1J_{CH}$ on changing from inert solvents, such as cyclohexane, to strongly donating ones, such as dimethyl methanamide and dimethyl sulphoxide. As may be seen, the values of $^1J_{CH}$ in iodomethane exhibit a much smaller solvent effect. Evans concludes that the nature of the interactions of $HCCl_3$ and H_3CI are rather different.

	$^1J_{CH}$/Hz	
	CCl_4 solvent	$HC(O)N(CH_3)_2$ solvent
$HCCl_3$	$208·4 \pm 0·3$	$217·4 \pm 0·3$
H_3CI	$150·7 \pm 0·2$	$151·5 \pm 0·2$

3.5.3 Thermodynamic Parameters

The determination of complex formation constants by gas–liquid chromatography was applied by Sheridan, Martire and Tewari [352] to the interactions of 1,1,1-trichloroethane with dioctyl ether and dioctyl sulphide. It was found that the formation constants were small, though far from insignificant, while the enthalpies and entropies of complex formation are comparable with those of dihalomethanes. Sheridan *et al.* point out that the technique of gas–liquid chromatography, with the donor as the stationary phase and the acceptor as the moving phase, is likely to measure a weighted average of 'pair-wise' interactions including not only hydrogen bonding, if present, but also a variety of donor–halogen interactions. From their data for this and other acceptors [352], they conclude that several interactions must be of importance since no simple relationship of thermodynamic parameters with molecular properties is apparent.

T/K	1,1,1-trichloroethane formation constants	
	$(C_8H_{17})_2O$	$(C_8H_{17})_2S$
$303·15$	$0·105$ dm^3 mol^{-1}	$0·165$ dm^3 mol^{-1}
$313·15$	$0·096$ dm^3 mol^{-1}	$0·152$ dm^3 mol^{-1}
$323·15$	$0·088$ dm^3 mol^{-1}	$0·140$ dm^3 mol^{-1}
$333·15$	$0·080$ dm^3 mol^{-1}	$0·129$ dm^3 mol^{-1}
$\Delta_\phi H$/kJ mol^{-1}	$-7·57$	$-7·07$
$\Delta_\phi S$/J K^{-1} mol^{-1}	$-43·7$	$-37·7$

3.5.4 Second Virial Coefficient

It has been found [233] that the second virial coefficients of ethanenitrile and propanone in the vapour phase do not agree with the values predicted by the Berthelot equation. The discrepancy is attributed to the formation of dimeric complexes, in which the energy of bonding is accounted for completely by dipole–dipole interaction.

the hypocrisies of marriage, which would have made her at least comprehensible to Aunt Vandalia, who was, like all Southerners, accustomed to the sexual duplicities of our traditional culture— Helen not only resisted conformity to the norms of her era, but also renounced the whole kit of beliefs about Southern womanhood. Had he lived to see it, the Captain would surely have been displeased by his youngest daughter's chosen way of living—though perhaps secretly gratified by the personal success and wealth she attained by leaving home and grasping life with both hands, making the most of her business talents in a great, distant city.

Perhaps—but not certainly. He smiled seldom, if ever, on the accomplishments of his children. Having endured much hardship and risk to create an upbringing for his children among deluxe commodities, pleasures, and smiles—rinsed free of Lost Cause nostalgia, firmly open to American mobility and enterprise—he was nevertheless always a stranger to this new world.

I was reminded of this fissure in the soul of my great-grandfather when I discovered, in Aunt Vandalia's house, a formal portrait of the Captain taken late in life. The portraitist has set his subject firmly against an unadorned backdrop, then captured the Captain's likeness as he wished to be seen: large of girth—his age's universal signal of worldly prosperity—solemn, attired in decidedly old-fashioned clothing, somewhat grim. It is an unrelaxed, self-conscious picture, more a monument to what he had made himself into than an intimate commemoration.

Despite this rigid artificiality—or perhaps because I know too well the brittle artifices one can adopt to ease displacement from one world of memory and imagination into another—I have treasured this image ever since the day I found it. And with the passage of years, I have come increasingly to honor the memory of the man, and respect him for his tenacity in negotiating the difficult passage from the Old South to the New. If indeed Captain Mays was the despot of family memory, disliked and avoided by his children, he is a kinsman I understand, perhaps more so now, in my own middle years. All his life, he was a man who belonged nowhere, a foreigner to the pleasures his offspring enjoyed. He portrayed himself as a New Southerner, but, in the photograph, the stern expression and old-fashioned suit and tie seem to encase him, like prison bars.

As I glance now at the portrait John Mays left—hanging just opposite me, as it has for years, in the room where I write this—the carefully accumulated trappings of age and authority, the symbolic paunch and jowls, the clenched jaw, the stiff collar and expensive, uncomfortable suit, all melt away. Left is a slender boy in a loose, tattered uniform where the old man was, only a moment ago. Our eyes meet across the distance of decades—this boy, keeping lonely watch on the Savannah River in 1865, and the lonely boy I was a century later, bearing the same ancient names passed down through centuries and generations—and we recognize each other.

Unsmiling, beloved ancestor: we are sons of the same Southern blood, you and I—the yield of different harvests from the broad, sun-washed fields of the South, and the loam of a single family. Our bones and sinews were knit in the womb of our native land, our minds and deepest understandings formed by its stories, its natural beauties, by the crimes and noble traditions of Southern civilization. Yet, more forcefully than by the blood we share, or even by the South that nursed us, we are joined in the maelstrom of time by our wounding, and our common exodus through the ceaseless unmaking and remaking of worlds, our common arrivals on distant shores, where we will always be strangers.

	CH$_3$CN	CH$_3$COCH$_3$
Observed binding energy/kJ mol^{-1}	21·8	13·4 to 33·5
Dipole–dipole energy/kJ mol^{-1}	26·3	13·9

The temperature dependence of the enthalpy of binding in propanone could not be accounted for; it is peculiar that this measured quantity should *increase* with increasing temperature.

3.5.5 Enthalpy of Mixing

It has been found [259] that nitromethane has a negative enthalpy of mixing with dimethyl ethanamide: $\Delta_m H(276\ \mathrm{K}) = -495$ J mol^{-1}. It would be interesting to be able to factor this result into a formation constant and an enthalpy of formation (cf. [7]).

3.5.6 NMR Solvent Shifts

Table 3.34 gives the change in chemical shift of H$_3$CX molecules on going from the vapour phase to solution in a variety of solvents. Interestingly, the shift changes are similar in all the solvents except diiodomethane and possibly tribromomethane. The shifts in all the solvents fall into the order

$$\mathrm{H_3CI} \approx \mathrm{H_3CCl} < \mathrm{H_3CBr} < \mathrm{H_3CCN}$$

Even in the strongly donor solvent dimethyl sulphoxide, the shifts are not in any way exceptional; hence, no hydrogen bonding interaction is indicated.

The solvent shifts of H$_3$CX compounds shown in Table 3.35 give little or no indication of an appreciable degree of complex formation by these compounds.

TABLE 3.1

Stoichiometry of higher complexes involving trichloromethane and an electron donor

	Donor	Complex	Method of detection	Ref.
(a) *Multiple donor sites*			
1.	ethyl ethanoate	A$_2$B	infrared spectroscopy	[205]
2.	butyl 2-methylpropanoate	A$_2$B	infrared spectroscopy	[83]
3.	dimethyl ethanamide	A$_2$B	infrared spectroscopy	[83]
4.	1-methylazolidinone-2	A$_2$B	infrared spectroscopy	[83]
5.	1,2-diethyl 1,2-diphenyl urea	A$_2$B	infrared spectroscopy	[83]
6.	tetramethyl urea	A$_2$B	infrared spectroscopy	[83]
7.	1,4-dioxane	A$_2$B	enthalpy of mixing	[265]
8.	1,4-dioxane	A$_2$B	polarisation	[18]
9.	7,14-dioxaeicosane	A$_2$B	enthalpy of mixing	[53]
10.	2,5,8,11,14-pentaoxapentadecane	A$_3$B	enthalpy of mixing	[406]
11.	1-phenyl-2,6,9-trioxa-5-aza-1-silatricyclo-[5.5.5.0^{1,5}]undecane	A$_2$B	dielectric loss	[74]

8

FESTIVE
TECHNOLOGY

Window, Henderson

MANY YEARS BEFORE I RECOVERED THE PORTRAIT OF Captain Mays from Aunt Vandalia's attic, and first recognized my face and history reflected in his, I knew the house at Greenwood sheltered a hoard of photographs.

On the infrequent afternoons during my childhood when Sister Erin allowed me some supervised rummaging in the shadowy cupboards, photos always turned up, heaped in cartons or pasted down on coarse black pages of large photograph albums, fragrant with age and the sweet decay of many Southern summers. Over many years, it seemed, Sister Erin and Aunt Vandalia had dumped new images on this accumulating heap of family pictures, only occasionally bothering to organize them into groups or albums. These dusty compilations were, in that sense, like the dolls Margaret and I had found in the attic—heirlooms cherished, but stuck away casually and hastily, with every good intention of coming back to them someday, to tidy them up and arrange them into some intelligible order. Neither my grandmother nor aunt ever got around to doing so.

That much I knew. But of the magnitude of this treasury, I knew very little. And before 1990, I had never wondered about the family stories that might be hidden in those photographs, waiting to be read. My first step on the path of memory was still a ways off, this hunt for traces of my family's passage through time and the Southern landscape still unimagined, even unimaginable. What I understood about our family photographs before Aunt Vandalia's death had come through visiting often and living briefly in the house in Greenwood, where such pictures were appreciated indifferently.

A few studio portraits, and one or two casual photographs enlarged and framed, hung on the walls of bedrooms and corridors as decoration, or stood propped on dressing tables and desks in exactly the same places, all my life. They were ceremonial pieces, for the most part, and almost never talked about. The largest formal portrait, installed over the piano, depicted Sister Erin's Alabama grandmother, whose beauty was legendary in our family, dressed like a

TABLE 3.1—*continued*

	Donor	Complex	Method of detection	Ref.
12.	1-ethoxy-2,6,9-trioxa-5-aza-1-silatricyclo-[5.5.5.0^{1,5}]undecane	A$_2$B	dielectric loss	[74]
	(b) *Single donor site*			
13.	cyclohexanone	A$_2$B	infrared spectroscopy	[396]
14.	propanone	A$_2$B	infrared spectroscopy	[396]
15.	propanone	A$_2$B	enthalpy of mixing	[211]
16.	propanone	A$_2$B	dielectric relaxation	[127]
17.	propanone	A$_2$B	polarisation [18],	[210]
18.	heptanone-4	A$_2$B	infrared spectroscopy	[104]
19.	bornanone	A$_2$B	infrared spectroscopy	[104]
20.	2,2,4,4-tetramethylpentanone-3	A$_2$B	infrared spectroscopy	[104]
21.	dimethyl sulphoxide	A$_2$B	NMR spectroscopy	[264]
		A$_2$B	freezing point diagram	[306]
22.	diethyl ether	A$_2$B, AB$_2$, AB$_3$	freezing point diagram	[403]
23.	oxolane	A$_2$B	polarisation	[393]
24.	dodecanol	A$_2$B/AB$_2$^{a}	polarisation	[254]
25.	tetraethylammonium bromide	A$_2$B	pressure/composition	[103]
26.	tetrabutylammonium bromide	A$_2$B	pressure/composition	[103]
27.	tetrabutylammonium chloride	A$_3$B	pressure/composition	[103]
28.	tetrabutylammonium bromide	A$_3$B	pressure/composition	[103]
29.	tetrabutylammonium iodide	A$_3$B	pressure/composition	[103]
30.	'propionopiperidide'	A$_2$B	infrared spectroscopy	[83]
31.	'isobutyropiperidide'	A$_2$B	infrared spectroscopy	[83]
32.	hexene-1	A$_2$B	NMR spectroscopy	[323]
33.	butanenitrile	A$_3$B	freezing point diagram	[277]

^{a} The experiment could not distinguish between A$_2$B and AB$_2$.

TABLE 3.2

Formation constants for 1:1 complexes of trichloromethane

Donor	T/K	K	Units^a	Solvent	Ref.
trichloromethane	298	0.16 ± 0.006	mf^{-1}	c-C$_6$H$_{12}$	[200]
	298	0.13 ± 0.03	mf^{-1}	c-C$_6$H$_{12}$	[180]
	301	0.0126 ± 0.0015	M^{-1}	c-C$_6$H$_{12}$	[400]
	298	0.13	M^{-1}	CCl$_4$	[359]
	313	0.10	M^{-1}	CCl$_4$	[359]
dichloromethane	293	0.07	M^{-1}	c-C$_6$H$_{12}$	[291]
1-iodobutane	–	0.6 ± 0.4	M^{-1}	–	[198]
	–	0.19 ± 0.02	M^{-1}	–	[56]
propanone	301·30	1.53	mf^{-1}	vapour	[29]
	313·55	1.33	mf^{-1}	vapour	[29]
	328·25	1.03	mf^{-1}	vapour	[29]
	303·15	1.59	mf^{-1}	vapour	[340]

^{a} M$\equiv$mol dm^{-3}; mf $\equiv$ mol mol^{-1} (mole fraction).
^{b} *cis*-decahydronaphthalene.

propanone–*cont.*	343	0·85	mf^{-1}	vapour	[340]
	363	0·57	mf^{-1}	vapour	[340]
	250	4·0 ± 2·0	mf^{-1}	--	[183]
	301	1·8 ± 0·6	mf^{-1}	–	[183]
	293	1·03	mf^{-1}	–	[18]
	293	0·35	mf^{-1}	–	[145]
	298	0·967	mf^{-1}	–	[211]
	298	4·1$_5$	mf^{-1}	–	[244]
	308	0·844	mf^{-1}	–	[211]
	323	0·698	mf^{-1}	–	[211]
	–	0·9	M^{-1}	–	[290]
	296	7·60 ± 0·05	mf^{-1}	c-C$_6$H$_{12}$	[59]
	218–300	1·2 ± 0·3	M^{-1}	c-C$_6$H$_{12}$	[396]
	301	0·751 ± 0·012	M^{-1}	c-C$_6$H$_{12}$	[400]
	304	0·90 ± 0·05	M^{-1}	C$_6$H$_{14}$	[203]
	298	0·8$_5$ ± 0·15	M^{-1}	C$_6$H$_{14}$	[54]
	298	0·8$_5$ ± 0·15	M^{-1}	C$_{10}$H$_{22}$	[54]
	298	0·8$_5$ ± 0·15	M^{-1}	C$_{14}$H$_{30}$	[54]
	298	0·8$_5$ ± 0·15	M^{-1}	C$_{16}$H$_{34}$	[54]
	298	0·3 ± 0·15	M^{-1}	CS$_2$	[54]
	298	0·4 ± 0·15	M^{-1}	CCl$_4$	[54]
	258	3·2 ± 0·2	mf^{-1}	CCl$_4$	[100]
	274	2·5 ± 0·2	mf^{-1}	CCl$_4$	[100]
	298	1·9 ± 0·1	mf^{-1}	CCl$_4$	[100]
	298	1·75 ± 0·04	mf^{-1}	CCl$_4$	[180]
	298	0·80	M^{-1}	?	[358]
	298	6·4	mf^{-1}	C$_6$H$_{14}$	[54]
	298	4·4	mf^{-1}	C$_{10}$H$_{22}$	[54]
	298	3·3	mf^{-1}	C$_{14}$H$_{30}$	[54]
	298	2·9	mf^{-1}	C$_{16}$H$_{34}$	[54]
	298		mf^{-1}	CS$_2$	[54]
	298	4·1	mf^{-1}	CCl$_4$	[54]
butanone	–	2·60	M^{-1}	–	[46]
cyclohexanone	–	2·20	M^{-1}	–	[46]
	300	1·2 ± 0·3	M^{-1}	c-C$_6$H$_{12}$	[396]
	301	1·023 ± 0·022	M^{-1}	c-C$_6$H$_{12}$	[400]
4-amino-1-acetylbenzene	308	0·14	M^{-1}	c-C$_6$H$_{12}$	[397]
benzaldehyde	–	1·5	M^{-1}	CCl$_4$	[226]
ethanoic acid	–	0·2	M^{-1}	–	[20]
methyl methanoate	323	0·0170	atm^{-1}	vapour	[232]
	350	0·0096	atm^{-1}	vapour	[232]
	368	0·0065	atm^{-1}	vapour	[232]
propyl methanoate	324	0·0341	atm^{-1}	vapour	[232]
	335	0·0256	atm^{-1}	vapour	[232]
	353	0·0158	atm^{-1}	vapour	[232]
	368	0·0114	atm^{-1}	vapour	[232]
methyl ethanoate	323	0·0280	atm^{-1}	vapour	[232]
	338	0·0247	atm^{-1}	vapour	[232]
	353	0·0210	atm^{-1}	vapour	[232]
	368	0·0192	atm^{-1}	vapour	[232]
ethyl ethanoate	323	0·0420	atm^{-1}	vapour	[232]
	338	0·0282	atm^{-1}	vapour	[232]
	353	0·0223	atm^{-1}	vapour	[232]
	368	0·0200	atm^{-1}	vapour	[232]
	301	0·674 ± 0·025	M^{-1}	c-C$_6$H$_{12}$	[400]
phenyl ethanoate	–	1·5	M^{-1}	CCl$_4$	[226]
trisethoxyphosphine oxide	293	1·26	M^{-1}	CCl$_4$	[156]
	308	1·10	M^{-1}	CCl$_4$	[156]

TABLE 3.2—*continued*

Donor	T/K	K	Units[a]	Solvent	Ref.
trisethoxyphosphine oxide —*cont.*	323	0·92	M^{-1}	CCl_4	[156]
	293	5·60	M^{-1}	$c\text{-}C_6H_{12}$	[156]
	301	4·637 ± 0·015	M^{-1}	$c\text{-}C_6H_{12}$	[400]
	308	4·17	M^{-1}	$c\text{-}C_6H_{12}$	[156]
	323	3·15	M^{-1}	$c\text{-}C_6H_{12}$	[156]
tributoxyphosphine oxide	283	7·0	M^{-1}	$c\text{-}C_6H_{12}$	[289]
	294	5·2	M^{-1}	$c\text{-}C_6H_{12}$	[289]
	308	3·8	M^{-1}	$c\text{-}C_6H_{12}$	[289]
	315	3·2	M^{-1}	$c\text{-}C_6H_{12}$	[289]
	—	1·0 ± 0·1	M^{-1}	CCl_4	[315]
diethoxymethylphosphine oxide	293	1·47	M^{-1}	CCl_4	[156]
	308	1·27	M^{-1}	CCl_4	[156]
	323	1·01	M^{-1}	CCl_4	[156]
diethoxy(2-propyl)phosphine oxide	293	1·41	M^{-1}	CCl_4	[156]
	308	1·20	M^{-1}	CCl_4	[156]
	323	0·98	M^{-1}	CCl_4	[156]
	293	6·13	M^{-1}	$c\text{-}C_6H_{12}$	[156]
	308	4·36	M^{-1}	$c\text{-}C_6H_{12}$	[156]
	323	3·23	M^{-1}	$c\text{-}C_6H_{12}$	[156]
diethoxy(trichloromethyl)- phosphine oxide	293	0·66	M^{-1}	CCl_4	[156]
	308	0·64	M^{-1}	CCl_4	[156]
	323	0·50	M^{-1}	CCl_4	[156]
	293	2·32	M^{-1}	$c\text{-}C_6H_{12}$	[156]
	308	1·83	M^{-1}	$c\text{-}C_6H_{12}$	[156]
	323	1·47	M^{-1}	$c\text{-}C_6H_{12}$	[156]
dibutoxybutylphosphine oxide	—	1·6 ± 0·2	M^{-1}	CCl_4	[315]
butoxydibutylphosphine oxide	—	2·4 ± 0·3	M^{-1}	CCl_4	[315]
tributylphosphine oxide	—	4·8 ± 0·3	M^{-1}	CCl_4	[315]
trioctylphosphine oxide	283	20·2	M^{-1}	$c\text{-}C_6H_{12}$	[289]
	294	13·9	M^{-1}	$c\text{-}C_6H_{12}$	[289]
	308	9·3	M^{-1}	$c\text{-}C_6H_{12}$	[289]
	315	7·5	M^{-1}	$c\text{-}C_6H_{12}$	[289]
triphenylphosphine oxide	293	2·28	M^{-1}	$c\text{-}C_6H_{12}$	[156]
	308	1·89	M^{-1}	$c\text{-}C_6H_{12}$	[156]
	323	1·52	M^{-1}	$c\text{-}C_6H_{12}$	[156]
tris(dimethylamino)phosphine oxide	293	3·02	M^{-1}	CCl_4	[156]
	308	2·36	M^{-1}	CCl_4	[156]
	323	1·98	M^{-1}	CCl_4	[156]
	293	13·36	M^{-1}	$c\text{-}C_6H_{12}$	[156]
	308	9·61	M^{-1}	$c\text{-}C_6H_{12}$	[156]
	323	6·99	M^{-1}	$c\text{-}C_6H_{12}$	[156]
	293	13·4	M^{-1}	$c\text{-}C_6H_{12}$	[291]
	293	0·07	M^{-1}	H_2CCl_2	[291]
trichlorophosphine oxide	283	8·8	mf^{-1}	—	[325]
	283	6·3	mf^{-1}	—	[326]
	368	4·0	mf^{-1}	—	[325]
	368	1·4	mf^{-1}	—	[326]
diethyl ether	298	0·13 ± 0·03	$mmHg^{-1}$	vapour	[75]
	308	0·11 ± 0·03	$mmHg^{-1}$	vapour	[75]

duchess in a black silk bodice with great puffs at the shoulder, that I always likened to angel's wings. There was also a dignified picture of my handsome grandfather John, who had brought his young wife from Texas to Louisiana around 1910, when the children were still toddlers. And prominently displayed over the writing desk in her room—it was always, for some reason, Sister Erin's favorite recollection of her son—was the blown-up snapshot of my father, kneeling in a knightly pose, rifle in hand, beside a wild ram he had just shot dead in the Sonora desert. A few smaller photographs were kept in an album bound in elaborately sculpted celluloid, and placed beside a massive family Bible on a conspicuous table in the living room.

Apart from the stiff professional photographs in this plump book or on the walls, and a few small paintings and too-sweet reproductions of popular devotional pictures—Dürer's praying hands, Holman Hunt's *Light of the World,* and such—art did not appear in the public areas of the house. Even the honored large portraits seemed cramped and out of place; for Sister Erin had not designed her dwelling place with art in mind. The outer walls were mostly glass, either in windows or doors; the inner ones were settings for fireplaces and bookcases and armoires. The room with the largest expanse of empty wall, my bedroom during the two years I lived at Greenwood, was a museum of heads lopped off the deer and other large animals my father shot on hunting trips in the 1930s. The photo album in the living room was not meant to be opened and perused. Like the assembly of stuffed heads and the enormous family Bible, and my grandmother's piano in its carved walnut case, this sumptuously decorated book was merely another standard fixture in the Edwardian era's inventory of domestic ornaments intended to impress visitors with the bourgeois respectability of the household.

The abundance of *other* family photographs—the snapshots secreted in many dusky corners—only began to dawn on me after I located them, and started to draw them from hiding places in closet darkness, chests, and trunks. As the unpacking of the house's stuffed niches progressed, more and more boxes and albums of pictures and negatives came to light. Spreading out a few images of my kin, picnicking or posing in new hats, swimming or holidaying, I sensed that I'd stumbled on an archive fascinating in ways I could not then put into words—more immediately arresting, surely, than the ancestral file, even more so than the dolls my wife and I found fast asleep in their little steamer trunks.

In the hectic days after Aunt Vandalia's death, however, emptying the house absorbed all our energies, and left no time to study anything we were finding. So along with my aunt's genealogical papers, the family china and silver, her stacks of Fiesta Ware and mementos and scrapbooks and much else, the photographs were quickly banished again into darkness, this time in newer cartons strapped shut for the long journey home to Toronto.

They might have stayed snug in their neat packing boxes much longer than they did had I not undergone another of those hauntings that started in Aunt Vandalia's attic, and would again summon me into the forests of family memory.

It all began with an invitation that came not long before my aunt's death in the spring of 1990. I had been asked to deliver a lecture on photography, one presentation in an annual series sponsored by Kodak at Toronto's Ryerson Polytechnic University. My speech was scheduled for early in the next year, and the choice of topic was left up to me. I was honored to be asked, and I accepted, though without having any idea of what I might have to say about photographic art. And no sooner had I begun puzzling that out, Aunt Vandalia died, and the job of closing her house—closing the chapter in the history that had been written there on post and lintel, flesh and blood, by two generations of my family's dwelling there—abruptly fell on me. That summer's work preempted concentration on almost everything else.

Suddenly, autumn was upon me, and I still did not have a topic for the Kodak lecture. A subject occasionally drifted up into my mind, only to sink, almost at once, back down into darkness. And the more seriously I pondered the matter, the less certain I became that I had anything to say. Which, given my career and history, came as a mildly troubling surprise.

I began to write about art in the early 1970s. Until almost the end of the previous decade, paint and bronze and steel comprised the stuff of top-deck American art, with photography as a whole in steerage, and pictures made with pop technologies like the Polaroid camera beneath mention. But by the time I left teaching in 1972 and

diethyl ether—*cont.*	318	0.10 ± 0.03	$mmHg^{-1}$	vapour	[75]
	328	0.07 ± 0.03	$mmHg^{-1}$	vapour	[75]
	326.4	0.214	–	vapour	[130]
	338.2	0.167_7	–	vapour	[130]
	351.2	0.104_7	–	vapour	[130]
	363.2	0.084_0	–	vapour	[130]
	–	0.80 ± 0.15	M^{-1}	–	[246]
	–	0.9 ± 0.4	M^{-1}	–	[198]
	–	0.33 ± 0.09	M^{-1}	–	[56]
	273	72	mf^{-1}	–	[250]
	283	48	mf^{-1}	–	[250]
	293	33	mf^{-1}	–	[250]
	298	5.6	mf^{-1}	–	[251]
	293	5.3	mf^{-1}	–	[326]
	293	0.40–0.66	mf^{-1}	–	[113]
	293	0.44	mf^{-1}	–	[145]
	298.2	1.93 ± 0.30	mf^{-1}	–	[76]
	343	1.8	mf^{-1}	–	[326]
	297	4.40 ± 0.05	mf^{-1}	$c\text{-}C_6H_{12}$	[59]
	298	3.76 ± 0.10	mf^{-1}	$c\text{-}C_6H_{12}$	[180]
	298	1.22 ± 0.04	mf^{-1}	CCl_4	[180]
ethyl propyl ether	298	0.11 ± 0.03	$mmHg^{-1}$	vapour	[75]
	308	0.10 ± 0.03	$mmHg^{-1}$	vapour	[75]
	318	0.09 ± 0.03	$mmHg^{-1}$	vapour	[75]
	328	0.06 ± 0.03	$mmHg^{-1}$	vapour	[75]
	298.2	1.45 ± 0.23	mf^{-1}	–	[76]
dipropyl ether	308	0.10 ± 0.03	$mmHg^{-1}$	vapour	[75]
	318	0.09 ± 0.03	$mmHg^{-1}$	vapour	[75]
	328	0.07 ± 0.03	$mmHg^{-1}$	vapour	[75]
	223	5.08	mf^{-1}	–	[15]
	256.7	1.91	mf^{-1}	–	[15]
	273	1.68	mf^{-1}	–	[15]
	298	1.2	mf^{-1}	–	[15]
	298.2	1.52 ± 0.24	mf^{-1}	–	[76]
	323	0.87	mf^{-1}	–	[15]
di(2-propyl)ether	308	0.10 ± 0.03	$mmHg^{-1}$	vapour	[75]
	318	0.08 ± 0.03	$mmHg^{-1}$	vapour	[75]
	328	0.08 ± 0.03	$mmHg^{-1}$	vapour	[75]
	338	0.05 ± 0.03	$mmHg^{-1}$	vapour	[75]
	273	3.42	mf^{-1}	–	[15]
	293	0.58–0.95	mf^{-1}	–	[113]
	293	0.62	mf^{-1}	–	[145]
	298	2.41	mf^{-1}	–	[15]
	298.2	2.85 ± 0.44	mf^{-1}	–	[76]
	323	1.50	mf^{-1}	–	[15]
	298	1.77 ± 0.08	mf^{-1}	CCl_4	[180]
ethyl butyl ether	308	0.10 ± 0.03	$mmHg^{-1}$	vapour	[75]
	318	0.10 ± 0.03	$mmHg^{-1}$	vapour	[75]
	328	0.07 ± 0.03	$mmHg^{-1}$	vapour	[75]
	298.2	1.39 ± 0.22	mf^{-1}	–	[76]
dibutyl ether	298.2	1.50 ± 0.23	mf^{-1}	–	[76]
	–	0.56	M^{-1}	–	[214]
	301	0.243 ± 0.015	M^{-1}	$c\text{-}C_6H_{12}$	[400]
di(2-methyl-2-propyl)ether	273	2.05	mf^{-1}	–	[44]
	298	1.33	mf^{-1}	–	[44]
	323	1.23	mf^{-1}	–	[44]
dipentyl ether	223	4.76	mf^{-1}	–	[44]
	248	1.43	mf^{-1}	–	[44]

began reviewing, many serious artists on the advancing edge of visual creation were deeply involved in photography—artists who, only a few years before, would surely have pursued their visions as painters and sculptors. Photography rapidly took hold; and by 1990—despite a resurgence of old-fashioned easel painting in Europe and North America in the early 1980s—photography was firmly entrenched at the very center of American artistic practice. How and why this happened are interesting matters outside the scope of this book. I am summarizing this complicated history, perhaps a little too handily, only to emphasize the curious fact occurring to me then: that if I did not know what to talk about at Ryerson, it was hardly because I had seen little photography. From my very beginnings in writing about art, I had been swimming in it.

The knot of the problem, I quickly realized, lay in the change that photography had undergone over the previous twenty years, and the corresponding change in me. The work that engaged me powerfully in the 1970s had been casual, intimate, much given to storytelling— low on production values, perhaps, but high on revelation. The artists I admired were exploring the erotic and social underworlds of the Modern city, its marginal people and sites, mass-cultural phe- nomena such as fashion and advertising and television—aspects of the ordinary, ambiguous urban life that had been excluded from the inner sanctum of high American art for most of my lifetime. The images made by artists who engaged my attention were not created for a market, for there was none, but generated by curiosity. And in these pictures, I was finding a way into the pulse, perversity, and complexity of human experience in the technological civilization to which I had come a stranger.

But as photography was quickly mainstreamed into the university art departments and museums, that early, raw spirit of investigation waned, and so did my interest. By 1989, the official sesquicentennial of photography's invention, my early excitement had largely evapo- rated. The newish photography seemed dulled down by a deftness at "reconciling avant-garde ambitions with the rewards of commercial- ism," as Susan Sontag calls this knack in her book *On Photography*. Traveling for my newspaper from one mammoth show of contempo- rary photographic art to another during that year, in Toronto and Montreal, New York and Paris, I found few images charged with the surprise and curiosity that had once enchanted me. The standard fare, instead, was cautious, dry illustration of the stylish cultural theories being hashed out in academic seminars—feminist, deconstructionist, psychoanalytical, Marxist, "transgressive" in the least interesting way. (As artists should have learned long ago, and apparently never will learn, art museums are the last places on earth the discontented masses ever look for revolutionary inspiration.)

The textbook, it seemed, had replaced the living world as the central subject of photography. And perhaps because I had come late and falteringly to the living world, I experienced the widespread abandonment of it by intelligent photographers as a peculiarly grievous loss. It would have been easy to talk about *that* in my Ryerson lecture. But the assignment I'd accepted was to speak on photography, not myself.

Then somewhere in this tense moment, hard up against the deadline my hosts at Ryerson had set for giving them a topic, I remembered there was a large body of photographs available to me, fortunately innocent of both my own neurotic investments (so it seemed) and the stuffy trendiness of art-school seminars. It was amateur family photography. I knew almost nothing about it, other than the fact that just about everybody does it. I had only rarely taken pictures, and never used my camera to photograph wife and kids and pets and the other subjects beloved by snapshooters. But I had a thousand old examples, in numerous cartons stacked on shelves in my Toronto apartment; and they would serve.

So I hauled down my photographs, then set to work sorting out and trying to grasp what these prints had meant to my kin who'd taken them, and what they might teach us now about the vast, incessant contemporary enterprise of making these much-loved, unimportant images.

The preparation of the lecture during the winter of 1990–1991 turned out to be even more appealing than I'd hoped, largely because it involved spending many hours with the unknown photographs Aunt Vandalia had bequeathed me. A few fell outside my immediate focus of interest. The earliest one, for example—a tintype, perhaps made at the time of the Civil War, depicting a young Confederate cadet, identity unknown, stuffed rather uncomfortably into

TABLE 3.2–*continued*

Donor	T/K	K	Units[a]	Solvent	Ref.
dipentyl ether–*cont.*	273	1·16	mf^{-1}	–	[44]
	298	1·02	mf^{-1}	–	[44]
	323	0·75	mf^{-1}	–	[44]
dioctyl ether	303·15	0·393 ± 0·010	M^{-1}	–	[352]
	313·15	0·350 ± 0·010	M^{-1}	–	[352]
	323·15	0·312 ± 0·010	M^{-1}	–	[352]
	333·15	0·275 ± 0·010	M^{-1}	–	[352]
di(2-chloroethyl) ether	293	0·09–1·18	mf^{-1}	–	[113]
	293	0·21	mf^{-1}	–	[14ɔ]
1,4-dioxane	323	1·11	mf^{-1}	vapour	[265]
	293	1·40	mf^{-1}	–	[18]
	–	1·90	M^{-1}	–	[46]
	313	0·50	M^{-1}	C_7H_{16}	[193]
	301	0·582 ± 0·013	M^{-1}	c-C_6H_{12}	[400]
oxane	296	4·30 ± 0·05	mf^{-1}	c-C_6H_{12}	[59]
oxolane	276·1	9·32	mf^{-1}	c-C_6H_{12}	[96]
	300	5·4$_4$	mf^{-1}	c-C_6H_{12}	[96]
	310·2	4·43	mf^{-1}	c-C_6H_{12}	[96]
	317·5	4·14	mf^{-1}	c-C_6H_{12}	[96]
	328·8	3·20	mf^{-1}	c-C_6H_{12}	[96]
	293	2·4	M^{-1}	40% c-C_6H_{12}	[393]
	293	4	M^{-1}	60% c-C_6H_{12}	[393]
oxetane	296	5·80 ± 0·05	mf^{-1}	c-C_6H_{12}	[59]
2-methyloxetane	299	5·90 ± 0·05	mf^{-1}	c-C_6H_{12}	[59]
oxirane	296	3·75 ± 0·05	mf^{-1}	c-C_6H_{12}	[59]
methyloxirane	295	4·80 ± 0·05	mf^{-1}	c-C_6H_{12}	[59]
1,3-dioxolanone-2	296	0·06$_0$	mf^{-1}	–	[264]
4-methyl-1,3-dioxolanone-2	296	0·08$_8$	mf^{-1}	–	[264]
dimethyl sulphoxide	293·15	1·50	mf^{-1}	–	[307]
	298·15	1·35	mf^{-1}	–	[307]
	308·15	1·20	mf^{-1}	–	[307]
	318·15	1·11	mf^{-1}	–	[307]
	328·15	0·95	mf^{-1}	–	[307]
	296	0·30 ± 0·01	M^{-1}	–	[264]
	298	1·06	M^{-1}	CCl_4	[350]
	293	1·25	M^{-1}	CCl_4	[157]
	308	1·09	M^{-1}	CCl_4	[157]
	323	0·95	M^{-1}	CCl_4	[157]
dibutyl sulphoxide	293	1·29	M^{-1}	CCl_4	[157]
	308	1·20	M^{-1}	CCl_4	[157]
	323	1·04	M^{-1}	CCl_4	[157]
di-(2-propyl) sulphoxide	293	1·43	M^{-1}	CCl_4	[157]
	308	1·26	M^{-1}	CCl_4	[157]
	323	1·09	M^{-1}	CCl_4	[157]
thiolane 1-oxide	293	1·22	M^{-1}	CCl_4	[157]
	308	1·10	M^{-1}	CCl_4	[157]
	323	0·94	M^{-i}	CCl_4	[157]
diethoxy sulphoxide	293	0·32	M^{-1}	CCl_4	[157]
	308	0·30	M^{-1}	CCl_4	[157]
	323	0·28	M^{-1}	CCl_4	[157]
azolidinone-2	273·7	1·6 ± 0·3	mf^{-1}	–	[44]

azolidinone-2−*cont.*	294·7	1·4 ± 0·3	mf^{-1}	−	[44]
	313·7	1·2 ± 0·3	mf^{-1}	−	[44]
	332·7	1·0 ± 0·4	mf^{-1}	−	[44]
1-methylazolidinone-2	301	3·222 ± 0·085	M^{-1}	$c\text{-}C_6H_{12}$	[400]
methyl ethanamide	254	0·42	M^{-1}	CCl_4	[376]
	272	0·44	M^{-1}	CCl_4	[376]
	289	0·46	M^{-1}	CCl_4	[376]
	309	0·48	M^{-1}	CCl_4	[376]
dimethyl methanamide	−	6	mf^{-1}	−	[209]
	293	0·91	M^{-1}	CCl_4	[157]
	308	0·82	M^{-1}	CCl_4	[157]
	323	0·71	M^{-1}	CCl_4	[157]
dimethyl ethanamide	−	5	mf^{-1}	−	[209]
	254	1·34	M^{-1}	CCl_4	[376]
	272	1·13	M^{-1}	CCl_4	[376]
	289	0·99	M^{-1}	CCl_4	[376]
	309	0·82	M^{-1}	CCl_4	[376]
	301	0·9	M^{-1}	CCl_4	[375]
	301	1·7	M^{-1}	$c\text{-}C_6H_{12}$	[375]
diethyl ethanamide	293	1·11	M^{-1}	CCl_4	[157]
	308	1·00	M^{-1}	CCl_4	[157]
	323	0·86	M^{-1}	CCl_4	[157]
1-ethanoylperhydroazine	293	1·06	M^{-1}	CCl_4	[157]
	308	0·94	M^{-1}	CCl_4	[157]
	323	0·82	M^{-1}	CCl_4	[157]
1-methyl-1,2-dihydro-azinone-2					
	293	1·25	M^{-1}	CCl_4	[157]
	308	1·10	M^{-1}	CCl_4	[157]
	323	0·97	M^{-1}	CCl_4	[157]
diethyl chloroethanamide	293	0·75	M^{-1}	CCl_4	[157]
	308	0·70	M^{-1}	CCl_4	[157]
	323	0·62	M^{-1}	CCl_4	[157]
triethylamine	250	12·0 ± 4·0	mf^{-1}	−	[183]
	301	3·0 ± 1·0	mf^{-1}	−	[183]
	298	4 ± 1	mf^{-1}	−	[258a]
	298	5	mf^{-1}	−	[251]
	298·15	4·7	mf^{-1}	−	[172]
	298	0·133	M^{-1}	−	[214]
	−	0·69	M^{-1}	−	[46]
	−	0·36 ± 0·02	M^{-1}	CCl_4	[19]
	298	4·70 ± 0·12	mf^{-1}	$c\text{-}C_6H_{12}$	[180]
	298	4·2 ± 0·2	mf^{-1}	$c\text{-}C_6H_{12}$	[94]
	308·2	3·50	mf^{-1}	$c\text{-}C_6H_{12}$	[94]
	316·7	3·10	mf^{-1}	$c\text{-}C_6H_{12}$	[94]
	332·2	2·10	mf^{-1}	$c\text{-}C_6H_{12}$	[94]
	347·7	1·60	mf^{-1}	$c\text{-}C_6H_{12}$	[94]
	275	0·792 ± 0·009	M^{-1}	$c\text{-}C_6H_{12}$	[400]
	301	0·425 ± 0·004	M^{-1}	$c\text{-}C_6H_{12}$	[400]
	314	0·321 ± 0·004	M^{-1}	$c\text{-}C_6H_{12}$	[400]
	331	0·229 ± 0·002	M^{-1}	$c\text{-}C_6H_{12}$	[400]
tributylamine	275	0·45	M^{-1}	$c\text{-}C_6H_{12}$	[289]
	283	0·36	M^{-1}	$c\text{-}C_6H_{12}$	[289]
	295	0·28	M^{-1}	$c\text{-}C_6H_{12}$	[289]
	309·7	0·22	M^{-1}	$c\text{-}C_6H_{12}$	[289]
trihexylamine	275	0·29	M^{-1}	$c\text{-}C_6H_{12}$	[289]
	283	0·23	M^{-1}	$c\text{-}C_6H_{12}$	[289]
	295	0·17	M^{-1}	$c\text{-}C_6H_{12}$	[289]
	309·7	0·13	M^{-1}	$c\text{-}C_6H_{12}$	[289]

TABLE 3.2–*continued*

Donor	T/K	K	Units[a]	Solvent	Ref.
trioctylamine	275	0·24	M^{-1}	c-C_6H_{12}	[289]
	283	0·20	M^{-1}	c-C_6H_{12}	[289]
	295	0·15	M^{-1}	c-C_6H_{12}	[289]
	306	0·13	M^{-1}	c-C_6H_{12}	[289]
cyclohexyldimethylamine	298	0·77 ± 0·02	M^{-1}	c-C_6H_{12}	[1]
	313	0·58 ± 0·02	M^{-1}	c-C_6H_{12}	[1]
cyclohexylmethylamine	298	0·94 ± 0·02	M^{-1}	c-C_6H_{12}	[1]
	313	0·69 ± 0·02	M^{-1}	c-C_6H_{12}	[1]
diethylamine	323	0·0390	atm^{-1}	vapour	[232]
	337	0·0297	atm^{-1}	vapour	[232]
dimethylamine	295	1·91	mf^{-1}	–	[58]
methylamine	295	3·45	mf^{-1}	–	[58]
propylamine	295	1·85	mf^{-1}	–	[58]
butylamine	295	2·44	mf^{-1}	–	[58]
cyclohexylamine	298	1·06 ± 0·02	M^{-1}	c-C_6H_{12}	[1]
	313	0·80 ± 0·02	M^{-1}	c-C_6H_{12}	[1]
	285·2	1·41	M^{-1}	c-C_6H_{12}	[397]
	290·7	1·31	M^{-1}	c-C_6H_{12}	[397]
	301·2	1·04	M^{-1}	c-C_6H_{12}	[397]
	303·2	1·03	M^{-1}	c-C_6H_{12}	[397]
	311·2	0·84	M^{-1}	c-C_6H_{12}	[397]
	315·7	0·84	M^{-1}	c-C_6H_{12}	[397]
	324·7	0·64	M^{-1}	c-C_6H_{12}	[397]
2-methyl-2-aminopropane	308	10	mf^{-1}	$HCCl_3$	[11]
aziridine	295	1·70	mf^{-1}	–	[58]
	253	7 ± 2	mf^{-1}	–	[331]
	273	3·8 ± 1	mf^{-1}	–	[331]
	311	3 ± 1	mf^{-1}	–	[331]
azetidine	295	2·20	mf^{-1}	–	[58]
azolidine	295	1·85	mf^{-1}	–	[58]
perhydroazine	295	2·05	mf^{-1}	–	[58]
perhydroazepine	295	2·20	mf^{-1}	–	[58]
1-methylazolidine	263·2	5·02	mf^{-1}	–	[36]
	280·7	3·85	mf^{-1}	–	[36]
	296·7	2·19	mf^{-1}	–	[36]
	296·7	2·2	mf^{-1}	–	[37]
	310	2·2	mf^{-1}	–	[35]
N-ethylidene(2-propyl)amine	293	5·2	mf^{-1}	–	[37]
	310	5·2	mf^{-1}	–	[35]
ethanenitrile	–	1·05	M^{-1}	–	[46]
	310	3·2	mf^{-1}	–	[35]
	298	0·91 ± 0·05	mf^{-1}	–	[180]
cyclohexyl cyanide	282·0	2·44	mf^{-1}	–	[36]
	287·7	2·14	mf^{-1}	–	[36]
	293·2	2·04	mf^{-1}	–	[36]
	293·2	1·9	mf^{-1}	–	[37]
benzenecarbonitrile	–	0·60	M^{-1}	–	[46]
1-azanaphthalene	293	0·44	mf^{-1}	–	[145]
azine	–	0·63	M^{-1}	–	[46]
	–	0·5	M^{-1}	–	[372]
	280·7	0·86	mf^{-1}	–	[36]
	290·7	0·73	mf^{-1}	–	[36]
	302·7	0·66	mf^{-1}	–	[36]

his grays—and the professional photographs done to commemorate the weddings and graduations of my east Texas people, from the last quarter of the nineteenth century through the first three decades of the twentieth.

The far greater number of photos—the array I was most keenly interested in—were snapshots taken between 1890, about the time instant cameras became available in east Texas, and the end of the 1940s. Many people appear in these photographs, though the pictures have only one subject: the large union of kinships created in 1904 by the marriage of Captain Mays's son John Matthew Mays, my grandfather, to Erin Estelle Bass of Palestine, another east Texas commercial center some sixty miles southwest of Henderson.

By 1904, the Mays and Bass families enjoyed much in common, though each had arrived at the same high position in life by different routes. The fathers of both John and Erin had firmly planted themselves in east Texas after 1865—the Captain, as a refugee from rural South Carolina's impoverishment by war; Matthew Belton Bass, as an ambitious cotton broker already accustomed to wealth in his native Alabama. Both men had prospered in trade during the postwar influx of old Confederate families into the Texas hill country. Once established in their respective communities, both the Captain and M. B. Bass had put up houses they thought appropriate to their status. Captain Mays built a sturdy, old-fashioned, decidedly Southern Neoclassical dwelling on a Henderson street corner, while, in Palestine, M. B. Bass confected a strongly angled, towered storybook mansion according to the Picturesque notions of style, achingly fashionable in the 1880s among Americans made wealthy by commerce.

In addition to money, a comfortable position in life, and imposing houses, the Mays and Bass families had another thing in common: an exuberant delight in taking pictures. As soon as the young children had gotten their hands on the new Kodak box cameras—introduced by George Eastman in 1888—and far into adulthood, they were snapping photographs of one another on any and every pretext, having them printed up, often in many copies, then circulating these images within their families and among their circles of friends.

When first leafing through the albums and separating loose images made in the decades on either side of 1900, this habit of picture-taking indicated nothing noteworthy about the specific inclinations or desires of my Texas families. Taking pictures, I imagined, was just something everybody did, and had always done. The first hint that these images might be trying to tell me more came at the end of a frustrating search for books about amateur picture-taking—not exactly a hot topic in the discussion of photography—when I discovered Pierre Bourdieu's *Photography: A Middle-brow Art*. It is the story of the distinguished French sociologist's investigation, in the late 1950s, into the ways ordinary French folk, in villages and small towns, used cameras—and into their idea about photography itself.

What Bourdieu found in France a half-century ago, as it turned out, was almost exactly true in Texas around the turn of the twentieth century. The instant camera was hardly the property of everyone, as I had supposed, but only of a certain stratum in the social stack. Both humble villagers and aristocrats disliked snapshots, and favored formal portraits done at commercial studios. The amateur picture, on the other hand, was adopted readily and unapologetically by people much like my relations in Texas—affluent and mobile, bourgeois and on the move socially and economically.

Like the ones Bourdieu studied, the photographers in my family snapped the goings and comings in the rich sensuous field of bourgeois family life, its festivals and anecdotes and passages. In its passage into the hands of my ancestors and relatives, the camera was transfigured from a solemn studio machine into what Bourdieu wonderfully dubbed "festive technology." A happy toddler, dressed in a charming camisole and hatted with a huge sombrero, wanders along a spacious verandah, and gets his picture taken. A crew of girls, my grandmother's classmates at the Cincinnati conservatory where she studied piano and singing in the 1890s, dress up as spooks for a Halloween party, and act spooky as Sister Erin clicks the shutter.

Uncle Will Bass, a young man on the verandah of relatives in San Antonio, glances up from his newspaper at Erin with a marvelously mischievous smile. Then Will takes the camera to capture young Erin, sweet in her billowing summer whites, sitting on the porch alongside their elderly grandfather, stern in his rocking chair—and on and on.

The courtship and marriage of John and Erin begins and continues always in front of cameras held by family and friends. In Henderson, my smiling young grandfather kneels by a woodpile with my

azine—*cont.*	310	0·69	mf^{-1}	–	[35]
	294·9	1·76	mf^{-1}	–	[125]
	313·2	1·32$_5$	mf^{-1}	–	[125]
	333·2	1·10	mf^{-1}	–	[125]
	298	1·61 ± 0·04	mf^{-1}	CCl$_4$	[180]
	300	0·38 ± 0·07	M^{-1}	CCl$_4$	[158]
aminobenzene	283	0·579	M^{-1}	c-C$_6$H$_{12}$	[397]
	294·8	0·526	M^{-1}	c-C$_6$H$_{12}$	[397]
	303·7	0·500	M^{-1}	c-C$_6$H$_{12}$	[397]
	314·8	0·448	M^{-1}	c-C$_6$H$_{12}$	[397]
	322·7	0·400	M^{-1}	c-C$_6$H$_{12}$	[397]
	298	0·33 ± 0·01	M^{-1}	c-C$_6$H$_{12}$	[1]
	313	0·29 ± 0·01	M^{-1}	c-C$_6$H$_{12}$	[1]
(ethylamino)benzene	308	0·23	M^{-1}	c-C$_6$H$_{12}$	[397]
(methylamino)benzene	298	0·27 ± 0·01	M^{-1}	c-C$_6$H$_{12}$	[1]
	308	0·25	M^{-1}	c-C$_6$H$_{12}$	[397]
	313	0·24$_5$ ± 0·01	M^{-1}	c-C$_6$H$_{12}$	[1]
(dimethylamino)benzene	298	0·25$_5$ ± 0·01	M^{-1}	c-C$_6$H$_{12}$	[1]
	313	0·23 ± 0·01	M^{-1}	c-C$_6$H$_{12}$	[1]
4-ethoxy-1-aminobenzene	308	0·65	M^{-1}	c-C$_6$H$_{12}$	[397]
4-chloro-1-aminobenzene	308	0·33	M^{-1}	c-C$_6$H$_{12}$	[397]
3-chloro-1-aminobenzene	308	0·25	M^{-1}	c-C$_6$H$_{12}$	[397]
2-chloro-1-aminobenzene	308	0·16	M^{-1}	c-C$_6$H$_{12}$	[397]
benzene	313	0·13	M^{-1}	C$_7$H$_{16}$	[193]
	298·7	1·30	mf^{-1}	c-C$_6$H$_{12}$	[94]
	308·3	1·20	mf^{-1}	c-C$_6$H$_{12}$	[94]
	316·7	1·08	mf^{-1}	c-C$_6$H$_{12}$	[94]
	332·2	0·93	mf^{-1}	c-C$_6$H$_{12}$	[94]
	347·7	0·82	mf^{-1}	c-C$_6$H$_{12}$	[94]
	306·6	1·40	mf^{-1}	c-C$_6$H$_{12}$	[178]
	306·6	1·68	mf^{-1}	C$_{10}$H$_{18}$^{b}	[178]
	306·6	1·89	mf^{-1}	(c-C$_6$H$_{11}$)$_2$	[178]
	306·6	1·67	mf^{-1}	C$_{14}$H$_{30}$	[178]
	306·6	1·69	mf^{-1}	C$_{16}$H$_{34}$	[178]
methylbenzene	313	0·17	M^{-1}	C$_7$H$_{16}$	[193]
1,4-dimethylbenzene	306·6	1·59	mf^{-1}	c-C$_6$H$_{12}$	[178]
1,4-diethylbenzene	306·6	1·19	mf^{-1}	c-C$_6$H$_{12}$	[178]
1,3,5-trimethylbenzene	313	0·21	M^{-1}	C$_7$H$_{16}$	[193]
	306·6	1·11	mf^{-1}	c-C$_6$H$_{12}$	[178]
1,3,5-triethylbenzene	306·6	1·74	mf^{-1}	c-C$_6$H$_{12}$	[178]
1,3,5-tris(2-propyl)benzene	306·6	1·92	mf^{-1}	c-C$_6$H$_{12}$	[178]
1-methylnaphthalene	313	0·22	M^{-1}	C$_7$H$_{16}$	[193]
dimethyl disulphide	301	0·170 ± 0·003	M^{-1}	c-C$_6$H$_{12}$	[400]
thiirane	302	3·80 ± 0·05	mf^{-1}	c-C$_6$H$_{12}$	[59]
thietane	299	4·10 ± 0·05	mf^{-1}	c-C$_6$H$_{12}$	[59]
thiolane	301	0·275 ± 0·004	M^{-1}	c-C$_6$H$_{12}$	[400]
	299	2·60 ± 0·05	mf^{-1}	c-C$_6$H$_{12}$	[59]
thiane	299	3·40 ± 0·05	mf^{-1}	c-C$_6$H$_{12}$	[59]
diethyl sulphide	301	0·219 ± 0·002	M^{-1}	c-C$_6$H$_{12}$	[400]
	302·7	3·40 ± 0·05	mf^{-1}	c-C$_6$H$_{12}$	[59]
dioctyl sulphide	303·15	0·418 ± 0·010	M^{-1}	–	[352]
	313·15	0·379 ± 0·010	M^{-1}	–	[352]
	323·15	0·345 ± 0·010	M^{-1}	–	[352]
	333·15	0·313 ± 0·010	M^{-1}	–	[352]
nitroethane	301	0·358 ± 0·014	M^{-1}	c-C$_6$H$_{12}$	[400]
cyclohexyl isocyanide	273·7	2·52	mf^{-1}	–	[36]
	282·4	2·32	mf^{-1}	–	[36]
	283·0	2·29	mf^{-1}	–	[36]

TABLE 3.2–*continued*

Donor	T/K	K	Units[a]	Solvent	Ref.
cyclohexyl isocyanide–*cont.*	294·9	1·90	mf^{-1}	–	[36]
	293·2	1·8	mf^{-1}	–	[37]
tetrabutylammonium chloride	300	2·51 ± 0·17	M^{-1}	CCl_4	[158]
	300	1·18 ± 0·05	M^{-1}	CH_3CN	[158]
tetrabutylammonium bromide	300	1·78 ± 0·08	M^{-1}	CCl_4	[158]
	300	0·73 ± 0·09	M^{-1}	CH_3CN	[158]
tetraheptylammonium iodide	300	1·07 ± 0·43	M^{-1}	CCl_4	[158]
tetrabutylammonium iodide	300	0·48 ± 0·05	M^{-1}	CH_3CN	[158]

TABLE 3.3

Formation constants for 2:1 complexes of trichloromethane

Donor	T/K	K	Units[a]	Solvent	Ref.
propanone	293	1·21	mf^{-2}	–	[18]
	298	1·117	mf^{-2}	–	[211]
	308	0·924	mf^{-2}	–	[211]
	323	0·668	mf^{-2}	–	[211]
	218–300	0·2₄	M^{-2}	c-C_6H_{12}	[396]
cyclohexanone	218–300	0·2₄	M^{-2}	c-C_6H_{12}	[396]
dimethyl sulphoxide	296	0·030	M^{-2}	–	[264]
	293·15	4·45	mf^{-2}	–	[307]
	298·15	4·40	mf^{-2}	–	[307]
	308·15	3·40	mf^{-2}	–	[307]
	318·15	2·55	mf^{-2}	–	[307]
	328·15	2·10	mf^{-2}	–	[307]
1,4-dioxane	293	1·74	mf^{-2}	–	[18]
	323·15	1·38	mf^{-2}	–	[265]
oxolane	293	4	M^{-2}	40% c-C_6H_{12}	[393]
	293	11·2	M^{-2}	60% c-C_6H_{12}	[393]

[a] $M \equiv$ mol dm^{-3}; mf $\equiv$ mol mol^{-1} (mole fraction).

TABLE 3.4

Minimum enthalpies of mixing of $HCCl_3$ with donor molecules[a]

Donor	T/K	$-\Delta_m H/J\ mol^{-1}$	Ref.
cyclohexane	288–93	636	[288]
tetrachloromethane	298	228	[288]
diethyl ether	273	2800	[231]
	276	6020[b]	[250]
	276	2720	[341]
	276	2930	[259]
	288	5480[b]	[250]
	297	4770[b]	[250]
	298	2500	[341]

[a] Mole fraction of trichloromethane = 0·5
[b] See Text.
[c] J cm^{-3}.

infant father, courting couples wave farewell to the camera as they depart for afternoon buggy rides in the Texas hills, uncles and cousins play baseball on a dusty sandlot, my Great-Grandmother Georgia stares grimly into the camera, with Captain Mays scowling impatiently beside her. The children of the rapidly expanding family gambol in the scanty summer grass of Palestine and Henderson, throw snowballs during a rare storm, receive their dinners *chez playhouse* from the hands of black servants, or sit alert for the camera, in goat-drawn wagons, grasping little whips and reins in their tiny fists.

As I reviewed these delightfully various amateur pictures of my kinship, only one thing remained constant: photography itself—the low, steady rustle of shutters in the background of every conversation, rite of baptism or marriage or funeral, meeting or game or sociable afternoon entertainment. *Photography itself*: not merely recording, but also creating a certain iconography of family events. Each click of the camera pinpointed a crossing of weft and warp in a large fabric of social connection—a meeting or parting, a party or solemnity taking place in a certain moment within the history of a given kindred and its encircling friendships. And with each click, a new picture was added to the unfolding narrative of my families. With this thought kept firmly at the center of my attention to these pictures, each print gradually lost its particularity, and flowed together into a continuous flow of information, a circulation of visual data across distances and across time—defining the invisible genealogical lines that linked the individual characters in this frieze of imagery, delineating the small-town social world these people moved within.

The lecture at Ryerson came and went. But in the days and weeks following, the photographs I had been contemplating did not quietly recede into the background of my mind. An obvious reason lay in the intrinsic interest I'd found in even the most unremarkable of these images. Almost everyone likes to look at family pictures—or at least the friends I showed my own family's photographs to. The most common questions they asked, and I asked myself, were: Who are they? What are they doing?—questions I could only occasionally answer with any certainty. But the most frequent comment from the others who perused the images had to do with how different these ancestors and relations seemed from people nowadays, and how curiously different the images themselves were, when compared to other kinds of photography.

These were intriguing observations and they prompted me to wonder what, exactly, comprised these differences?

The answer, I decided, was to be found in the mere happiness pervading the imagery my family's photographers had created, the warm intimacy in which all alienation and suffering and bewilderment is absent. The old people of the Bass family, and Captain Mays, are everlasting, monumental presences in these pictures, firmly enthroned in their rockers or on voluminously stuffed sofas, their high collars propping up wise heads. The children of the Captain and Georgia Trammell—then later, the children of these sons and daughters, now grown up and married—are always delighted, frolicking on broad lawns dappled by shade trees and Texas sunshine, or trussed up in fancy costumes, or firmly in control of the pony they sit on. The older women are seated among their possessions, surrounded by progeny; the older men are at leisure on their verandahs. My grandfather and father as lads are stalwart, dressed smartly and posing easily for the camera; for they had grown up with this technology, and were comfortable on either side of it.

It was while gazing at these photographs, some time after my lecture, that I felt my interest in them as evidence gradually giving way to a deeper, ominous longing. It was an old desire, largely unfelt since the weight of too much yearning for the past precipitated the collapse decades before. It was the yearning to take leave of the present world, in which I had always been an alien and stranger, and disappear into the lost world described by those photographs, and into the happy east Texas families they represented with much tenderness. There, my kin drifted on a sea of pleasures, babies taken for horseback rides by the elders, toddlers kissing, strong boys boxing, my splendid uncles and cousins standing under the porticos of houses they had lived in all their lives. In these images I seemed to see the fullness of joy I had lost when my father died, when my mother and I left Spring Ridge—the *essence* of general tranquillity, nurtured in the generation of Captain Mays's children, carried among them into the twentieth century, and lingering still in the house at Greenwood in 1947, when the news of my father's death brought all crashing down.

diethyl ether—*cont.*	298	2500	[342]
	298	2634	[25]
	298	600^c	[75]
	308	510^c	[75]
	318	470^c	[75]
	328	330^c	[75]
ethyl propyl ether	298	510^c	[75]
	308	470^c	[75]
	318	420^c	[75]
	328	290^c	[75]
	298·2	2300	[76]
dipropyl ether	298	1942	[25]
	298·2	1940	[76]
	308	470^c	[75]
	318	420^c	[75]
	328	330^c	[75]
di(2-propyl)ether	276	3100	[341]
	298	2987	[25]
	298	2840	[341]
	298·2	2900	[76]
	308	470^c	[75]
	318	380^c	[75]
	328	380^c	[75]
	338	240^c	[75]
ethyl butyl ether	298·2	2220	[76]
	308	470^c	[75]
	318	470^c	[75]
	328	330^c	[75]
dibutyl ether	276	2010	[341]
	298	1720	[341]
	298	1816	[25]
	298·2	1810	[76]
di(2-methyl-2-propyl) ether	298	1520	[25]
dipentyl ether	298	1753	[25]
di(3-methylbutyl) ether	298	1441	[25]
dihexyl ether	298	1585	[25]
cyclohexyl methyl ether	276	2780	[259]
phenyl methyl ether	276	730	[259]
1,2-dimethoxyethane	276	5020	[406]
1-methoxy-1,1,2-trifluoro-2- chloroethane	273	389 ± 13	[231]
1-ethoxy-1,1,2-trifluoro-2- chloroethane	273	259 ± 17	[231]
1-propoxy-1,1,2-trifluoro-2- chloroethane	273	218 ± 17	[231]
1,4-dioxane	298	1900	[341]
	323·15	1900	[265]
oxane	276	2680	[343]
	276	2680	[341]
	298	2510	[341]
1-methyl-4-(2-propyl)-1,8- epoxycyclohexane (cineole)	298	3490	[341]
oxolane	276	3140	[343]
	276	3140	[341]
	298	2830	[341]
	298	2809	[276]
2,2-dimethyloxolane	298	3010	[341]
oxetane	276	3180	[343]

TABLE 3.4–*continued*

Donor	T/K	$-\Delta_m H/\text{J mol}^{-1}$	Ref.
oxetane–*cont.*	276	3180	[341]
	298	2940	[341]
2,2-dimethyloxetane	276	3830	[341]
2,2-diethyloxetane	298	3540	[341]
2-(chloromethyl)oxirane	276	770	[343]
	298	790	[341]
oxirane	276	1530	[343]
2-methyloxirane	276	1970	[343]
	276	1930	[341]
	298	1850	[341]
2-phenyloxirane	276	1030	[343]
	298	900	[341]
7-oxabicyclo[4.1.0]heptane	276	2790	[343]
	298	2370	[341]
poly(1,2-dimethoxyethane)	278·68	$49·8^c$	[253]
poly(2-methyloxirane)	287·68	$32·84^c$	[213]
benzyl ethanoate	298	$1230(\text{A}_2\text{B})$	[270]
methyl ethanoate	298	1550	[342]
ethyl ethanoate	298	1940	[342]
methyl propanoate	298	1820	[342]
3-propanolide	298	700	[342]
4-butanolide	298	1590	[342]
5-pentanolide	298	2360	[342]
propanoyl chloride	276	750	[14]
propanone	–	1970	[66]
	273	1970	[231]
	276	2070	[259]
	298	1760	[342]
	298	1905	[211]
	308	1920	[211]
	323	1805	[211]
butanone	276	2275	[259]
cyclohexanone	298	2410	[342]
3,5-dioxaheptanone-4	276	2190	[14]
nitromethane	276	0	[259]
nitrobenzene	276	495	[259]
ethyl nitrate	276	475	[259]
dimethyl sulphoxide	298·15	$3000(\text{A}_2\text{B})$	[121]
diethoxy sulphoxide	276	2110	[14]
diethoxy sulphone	276	940	[14]
	276	1075	[259]
ethyl benzenesulphonate	276	$1190(\text{A}_2\text{B})$	[259]
trichlorophosphine oxide	–	1540	[224]
	276	2020	[14]
dichloroethoxyphosphine oxide	–	1560	[224]
dichlorobutoxyphosphine oxide	–	1840	[224]
chlorodiethoxyphosphine oxide	–	2730	[224]
chlorodibutoxyphosphine oxide	–	2970	[224]
triethoxyphosphine oxide	–	4460	[224]
tripropoxyphosphine oxide	–	4710	[224]
tributoxyphosphine oxide	–	4650	[224]
	298	$5650(\text{A}_2\text{B})$	[1a]
tris(2-propoxy)phosphine oxide	–	4810	[224]

With these feelings came others, fresh stirrings of revulsion toward the official photographic art of museums and galleries that my job as a critic had always required me to see, and write about. What intellectual pleasure I had known, when reading Sontag's aphorisms and observations in her book *On Photography* and other meditations like it, vanished before the poignancy in these infinitely more precious pictures.

"To photograph is to appropriate the thing photographed," Sontag writes. But in the practice of my family's photographers, there was no raw, unfilled psychic distance between the photographer and those photographed, no need to take back and hold what had been lost, or was in danger of being lost. The exchange of glance between the photographed and the photographer flashes across a comfortable interval, suffused with affection. Eyes met. They did not appropriate.

Elsewhere she says, plausibly enough: "Photographs furnish evidence. Something we hear about, but doubt, seems proven when we're shown a photograph of it." But in common with all recent writers about photography, Sontag assumes a preexisting condition of alienation that the photograph somehow heals, a doubt that it can allay. In the evocative exchanges flickering among the photos that my family produced, and in the exchange between my own eyes and those of my long-dead kindred, I felt no trace of doubt, no hint of ignorance, no inner urgency on the part of the photographers to *prove* anything. The images celebrated the joy blooming in the daily lives of people who knew and loved each other.

And if the pictures refused to be high "photographic art"—and in fact, to my mind, condemned such art to irrelevance—they also resisted integration into the more popular, practical tasks of photographs, as aids to erotic arousal or religious devotion, providers of patriotic inspiration, information about new products, and so on. They did not appear to belong anywhere among the defining objects of everyday life in the Modern world of the late twentieth century—our appliances and apparel, newspapers and television, the imagery and language of mass culture, all ceaselessly producing in us the false notion of snug belonging in mass culture that Michel Foucault has called our civilization's "general politics of truth."

The family photographs were rich in a *specific* politics of truth, in knowledge of being in the world in certain places—Henderson,

Palestine, San Antonio, then Greenwood and Shreveport—among certain people in the web of intimate fixities and certainties described by genealogy. Their truth, sensuous and ambiguous and rich, arose from the pleasures and spectacles they recorded, and from the very absence of staging, careful lighting, and so forth thought de rigueur in the serious camerawork of my own time. Among the photographers whose practices he studied, Bourdieu found that the ones most preoccupied by the pedantry and laborious artifice of composition were precisely those whose roots in community and in the traditional loyalties of family life had been disturbed—*Moderns*, that is, in the full twentieth-century sense. The aficionados in my Texas families, from the 1890s into the adulthoods of such eager photographers as my father and Aunt Vandalia, did not feel the need to compose carefully. Genealogy had provided them with the composition that counted, assembling everything within the borders of these prints—the houses and children, the lawns and streets, clowning kids and serious men and women—into a radiant unity that all the photographic gadgetry and skill on earth could not create.

But in this moment of intense, haunted thought, when looking at my legacy of images and yearning for the peace I believed I saw in them, genealogy became the force that violated the fantasy, and kept me from falling into the poisoned waters of nostalgia. What gently tugged my thinking back on a sensible course were the scraps of family story festooning the immutable genealogical lines between the men and women, between generations and centuries of my kinship. Truths were coming back to mind as I looked at the photographs, and draining the images of their seductive powers.

Among the pictures was one of Aunt Vandalia, slender in a stylishly brief, waistless frock of the late 1920s, smiling discreetly through a lacy veil at the camera held by her brother. She is an untroubled and flowering woman, new to married life. Absent from the picture was what I knew (from diaries, from surmise) of her inner torment—the pangs of conscience she felt because she had married a divorced man, the worries over eyesight and general health that were already dogging her.

The next moment, my aunt takes the camera from my father, and points it. The tall boy pulls himself up and grins from under his

ethyldiethoxyphosphine oxide	–	4830	[224]
propyldiethoxyphosphine oxide	–	5160	[224]
butyldibutoxyphosphine oxide	–	5770	[224]
dibutylbutoxyphosphine oxide	–	5980	[224]
diethoxyphosphine oxide	–	3510	[224]
dibutoxyphosphine oxide	–	4100	[224]
tributoxyphosphine oxide	–	2850	[224]
dimethyl ethanamide	276	3850	[86]
	276	3850	[259]
diethyl benzamide	276	4560	[259]
butylamine	276	2930	[259]
	298	2990	[378]
cyclohexylamine	276	3345	[259]
di(2-propyl)amine	298	3910	[378]
diethylamine	298	3710	[378]
dibutylamine	298	3390	[378]
cyclohexyldimethylamine	276	4880	[259]
triethylamine	298	3640	[378]
	298·15	4030	[172]
	308·15	3600	[172]
benzenecarbonitrile	276	1245	[259]
pentanenitrile	276	1630	[259]
pentanedinitrile	276	1088(A₂B)	[259]
azole	298	605	[378]
2-azanaphthalene	298	1840	[378]
1-azanaphthalene	298	2030	[378]
azine	298	2025	[378]
	298	2155	[125]
2-methylazine	298	2470	[378]
3-methylazine	298	2240	[378]
4-methylazine	298	2330	[378]
2,4-dimethylazine	298	3230	[378]
2,6-dimethylazine	298	3250	[378]
2,4,6-trimethylazine	298	3780	[378]
oxole	298	212·5	[276]
2-methyloxole	298	440	[276]
benzene	288–93	375	[21]
	298	375	[377]
methylbenzene	298	655	[377]
1,2-dimethylbenzene	298	865	[377]
1,3-dimethylbenzene	298	850	[377]
1,4-dimethylbenzene	288–93	920	[21]
	298	845	[377]
1,3,5-trimethylbenzene	298	905	[377]
(dimethylamino)benzene	276	1170	[259]
1-methyl-2-chlorobenzene	298	155	[377]
chlorobenzene	288–93	175	[21]
	298	120	[377]
1,2-dichlorobenzene	298	0	[377]
1-chloronaphthalene	298	155	[377]
bromobenzene	298	140	[377]

a Mole fraction of trichloromethane = 0·5

b See Text.

c J cm^{-3}.

TABLE 3.5

Enthalpies of formation of trichloromethane complexes, calculated from equation 3.8

Donor	K/mf^{-1}[a]	$\Delta_m H/\text{kJ mol}^{-1}$	$\Delta_\phi H/\text{kJ mol}^{-1}$ Calculated
diethyl ether	5·3	−2·5	− 9·2
dipropyl ether	1·2	−1·9$_4$	−13·1
di(2-propyl) ether	2·41	−2·8$_4$	−13·7
di(2-methyl-2-propyl) ether	1·33	−1·5$_2$	− 9·7
dipentyl ether	1·02	−1·4$_4$	−10·6
1,4-dioxane	1·40	−1·9$_0$	−11·8
propanone	0·967	−1·7$_6$	−13·4
azine	1·76	−2·0$_3$	−11·3
triethylamine	4	−3·6$_4$	−14·5
butylamine	2·44	−2·9$_9$	−14·4

[a] mf ≡ mol mol^{-1} (mole fraction).

TABLE 3.6

Enthalpies and entropies of formation of trichloromethane complexes

Donor	T/K	$\Delta_\phi H/\text{kJ mol}^{-1}$	$\dfrac{\Delta_\phi S}{\text{J K}^{-1}\text{mol}^{-1}}$	Solvent	Ref.
propanone	250–301	−10·5 $\begin{cases} +10\cdot5 \\ -5\cdot3 \end{cases}$		−	[183]
	−	−17·0		−	[267]
	−	−11·3 ± 0·4		−	[66]
	298	−10·1		−	[211]
	308	−10·2		−	[211]
	323	−10·3		−	[211]
	−	− 4·2		−	[17]
	258–298	− 8·8 ± 1·7		CCl$_4$	[100]
	304	−14·6		C$_6$H$_{14}$	[203]
	298	− 9·8 ± 0·5	−35·0 ± 2·0	c-C$_6$H$_{12}$	[400]
butanone	−	− 5·4		−	[17]
3-methylbutanone-2	−	− 4·6		−	[17]
3,3-dimethylbutanone-2	−	− 3·4		−	[17]
cyclohexanone	300	−10·2 ± 0·4	−33·9 ± 1·5	c-C$_6$H$_{12}$	[400]
diphenyl ketone	−	− 7·3$_2$		−	[17]
methyl methanoate	323–368	−21·1	−99	vapour	[232]
propyl methanoate	324–368	−23·5	−101	vapour	[232]
methyl ethanoate	323–368	− 7·9	−54	vapour	[232]
ethyl ethanoate	296·7	−10·5 ± 0·4	−37·8 ± 1·3	c-C$_6$H$_{12}$	[400]
	323–368	−16·2	−77	vapour	[232]
diethyl ether	298·2	−13·2 ± 0·4		−	[76]
	293–343	−18·0 ± 4·2		−	[326]
	273–293	−25·4		−	[250]
	326·4–363·2	−25·2	−109	vapour	[130]
ethyl propyl ether	298·2	−12·7 ± 0·4		−	[76]
dipropyl ether	298	−10·8		−	[15]
	298·2	−10·4 ± 0·3		−	[76]
di(2-propyl) ether	298	−11·7		−	[15]
	298·2	−11·8 ± 0·4		−	[76]

[a] For 2:1 complex.

jaunty straw hat—and the shutter clicks. He smiles forever there, in the timeless time and place of photography. But absent is any trace of the anguish already stirring within—the ungrounded distraction that was never to allow him a certain, settled place, his gradual bowing and bending toward prolonged despair and alcoholic oblivion, that he would not escape until his destruction, one August night in 1947, on a lonely road in Oklahoma.

There was a picture of an elegant city home in south Texas, aloof in the hot summer air, embraced by luxurious plantings of palms. Written in white ink on the black album page was the legend: "House of M. B. Bass." Who, but one who remembered the stories told in our family, could know that the picture was taken years after Sister Erin's father had lost it, and all else, in the financial disaster that ruined him?

But if the pictures did not tell everything, was all the joy and pleasure in them *false*? Were the stories in these photographs *lies*? No, I decided; there were no lies in this compendium—only facts like those in fiction, true voices from the past that we can hear only across the passionate, distorting distances of time. As Roland Barthes jotted down at the beginning of his autobiography and album of family photos: "It must all be considered as if spoken by a character in a novel."*

The family snapshots that engrossed me were like such speeches—vignettes in a fiction without beginning or end, a new phrase or paragraph added each time the camera clicked. In common with all fictions, of course, my family photographs presented only a residue after all else that is known, but undesirable, is excluded—a process that leaves only *what fits*, enframed by four straight edges and surrounded by silences and unknowing, and stories known in tellings but not in seeing.

Once I acknowledged this ineradicable mythmaking, the family photographs blossomed with new meaning, opening pathways to the ideological fountains that watered the mentality of my kin in east Texas around the turn of the twentieth century.

My grandfather and Sister Erin, their siblings and kin, all loved to take pictures of children, for instance, but only did so when the children fit. And *fitting* had nothing to do with being cute or cuddly. My Texas kin knew nothing of what we call "baby pictures"—little por-

* Roland Barthes, tr. Richard Howard. *Roland Barthes* (New York: Hill and Wang, 1977).

traits of our beloved children and godchildren taken against bare backdrops. Little Vandalia and John, when babies and children, were always photographed as wise, serious children in luxurious environments, doing appropriately bourgeois things: playing on the strand while on a seaside holiday, a black maid close at hand, or frolicking on spacious, columned verandahs or in wide yards, or in other architectural settings that confirmed the station of John and Sister Erin in society. These children were never snapped naked, or while being stubborn or vulgar, or otherwise playing a fractious role within the general assonance of life that our family photography insisted on. The only kinds of children qualifying for photographic celebration were ones who, in that instant, expressed the ideal of childhood held dear in my bourgeois family at the turn of the century: plumply healthy, valiant, thoughtful, curious—always curious, and always content within the universe of commodities.

These commodities, and all significant possessions—houses, gins, mules, servants—were also considered worthy of photography. A pile of cotton in a field after harvest was acceptable; the same pile used as a backdrop to depict the field hands who'd harvested it was even desirable, because it asserted the mastery of the strong white hand over both land and blacks. (It would take American photographers, working in the 1930s, to come up with the idea of depicting the field hands in *their* social contexts, of poverty, life in shacks, on the margins of white culture.)

If family friends were often photographed, abandoned wives were not; and certainly not mistresses, despite their close proximity to everything the family did. The white woman with whom my grandfather John had a long, apparently happy affair was not, of course, invited to the wedding of his son to my mother in 1927. He arranged to meet this lady friend immediately after the ceremony and celebrations, and drove off with her for a brief holiday in the resort town of Hot Springs, Arkansas—only to run into my parents, who were honeymooning in the same town and the same hotel. While the encounter was remembered as funny—at least to my surprised father and grandfather—no photographic record of the woman exists, and even her name has not come down to me: for, if tolerated on the fringe of kinship's net, she was not permitted any status within it.

oxolane	298·2	−11·97		c-C_6H_{12}	[276]
	300	−15·1		c-C_6H_{12}	[96]
ethyl butyl ether	298·2	−12·5 ± 0·4		−	[76]
dibutyl ether	298·2	− 9·7 ± 0·3		−	[76]
	305	− 9·8 ± 0·5	−22·3 ± 0·8	c-C_6H_{12}	[400]
di(2-methyl-2-propyl)ether	298	− 7·9		−	[15]
dipentyl ether	298	−10·0		−	[15]
dioctyl ether	303·15–333·15	− 9·96	−40·5	−	[352]
1,4-dioxane	294	−10·7 ± 0·4	−40·3 ± 1·3	c-C_6H_{12}	[400]
	313	− 9·2		C_7H_{16}	[193]
	293	− 7·5		−	[13]
	323·15	− 8·4		−	[265]
trichlorophosphine oxide	283–368	− 8·4 ± 2·1		−	[325]
	283–368	−15·5 ± 4·2		−	[326]
triethoxyphosphine oxide	308	− 8·4	−25·9	CCl_4	[156]
	308	−15·1	−37·2	c-C_6H_{12}	[156]
	305	−15·9 ± 0·2	−40·2 ± 0·5	c-C_6H_{12}	[400]
tributoxyphosphine oxide	−	− 3·8		CCl_4	[315]
	294	−18·0	−47·3	c-C_6H_{12}	[289]
methyldiethoxyphosphine oxide	308	− 8·8	−26·8	CCl_4	[156]
2-propyldiethoxyphosphine oxide	308	− 9·6	−29·7	CCl_4	[156]
	308	−16·7	−42·3	c-C_6H_{12}	[156]
trichloromethyldiethoxy-phosphine oxide	308	− 7·1	−27·6	CCl_4	[156]
	308	−12·1	−33·9	c-C_6H_{12}	[156]
butyldibutoxyphosphine oxide	−	− 4·6		CCl_4	[315]
dibutylbutoxyphosphine oxide	−	− 5·9		CCl_4	[315]
tributylphosphine oxide	−	− 7·5		CCl_4	[315]
trioctylphosphine oxide	294	−20·9	−49·4	c-C_6H_{12}	[289]
triphenylphosphine oxide	308	−10·5	−29·3	CCl_4	[156]
tris(dimethylamino)phosphine oxide	308	−11·3	−28·9	CCl_4	[156]
	308	−17·2	−36·4	c-C_6H_{12}	[156]
dimethyl sulphoxide	298	− 6·7		CCl_4	[350]
	293·15–328·15	−16·3[a]		−	[307]
	293·15–328·15	− 9·8		−	[307]
	293–323	− 6·7	−22·6	CCl_4	[157]
dibutyl sulphoxide	293–323	− 7·5	−23·8	CCl_4	[157]
di(2-propyl) sulphoxide	293–323	− 7·9	−25·1	CCl_4	[157]
thiolane 1-oxide	293–323	− 7·1	−23·0	CCl_4	[157]
diethoxy sulphoxide	293–323	− 3·8	−17·2	CCl_4	[157]
'nitroalkanes'	−	− 1·67		−	[17]
nitroethane	315	− 6·4 ± 0·1	−29·8 ± 0·4	c-C_6H_{12}	[400]
azolidinone-2	294·7	− 6·0 ± 2		−	[44]
1-methylazolidinone-2	305	−16·69 ± 0·08	−45·6 ± 0·4	c-C_6H_{12}	[400]
ethyl ethanamide	−	− 5·0		−	[218]
methyl ethanamide	289	+ 1·7	− 0·8	CCl_4	[376]
dimethyl methanamide	293–323	− 5·9	−21·3	CCl_4	[157]
dimethyl ethanamide	289	− 4·6	− 15·5	CCl_4	[376]
diethyl ethanamide	293–323	− 6·7	−21·8	CCl_4	[157]
1-ethanoylperhydroazine	293–323	− 6·3	−22·2	CCl_4	[157]
1-methyl-1,2-dihydroazinone-2	293–323	− 6·7	−23·0	CCl_4	[157]
diethyl chloroethanamide	293–323	− 5·0	−20·5	CCl_4	[157]
triethylamine	305	−16·9 ± 0·13	−63·6 ± 0·4	c-C_6H_{12}	[400]
	250–301	−16·7 $\begin{cases} +16·7 \\ − 8·4 \end{cases}$		−	[183]
	298	−16·7		−	[214]
	298·15	−14·2	−35	−	[172]

TABLE 3.6–*continued*

Donor	T/K	$\Delta_\phi H/\text{kJ mol}^{-1}$	$\dfrac{\Delta_\phi S}{\text{J K}^{-1}\text{mol}^{-1}}$	Solvent	Ref.
triethylamine–*cont.*	298	$-17\cdot4 \pm 0\cdot8$	$-46\cdot0$	c-C_6H_{12}	[94]
tributylamine	294	$-15\cdot1$	$-61\cdot9$	c-C_6H_{12}	[289]
trihexylamine	294	$-15\cdot1$	$-65\cdot3$	c-C_6H_{12}	[289]
trioctylamine	294	$-14\cdot2$	$-64\cdot0$	c-C_6H_{12}	[289]
cyclohexyldimethylamine	298	$-16\cdot4 \pm 0\cdot2$		–	[1]
cyclohexylmethylamine	298	$-16\cdot4 \pm 0\cdot2$		–	[1]
diethylamine	323–337	$-17\cdot5$	-81	vapour	[232]
cyclohexylamine	298	$-16\cdot1 \pm 0\cdot2$		–	[1]
	298	$-15\cdot1 \pm 1\cdot3$		c-C_6H_{12}	[397]
1-methylazolidine	296·7	$-16\cdot3$		–	[36]
aziridine	273	$-11\cdot7 \pm 6\cdot3$		–	[331]
azine	290·7	$-8\cdot8$		–	[36]
	294·9–333·2	$-10\cdot1 \pm 1\cdot1$		–	[125]
cyclohexyl cyanide	293·2	$-10\cdot9$		–	[36]
azobenzene	–	$-2\cdot5$		–	[17]
aminobenzene	298–308	$-7\cdot1 \pm 0\cdot8$		c-C_6H_{12}	[397]
	298	$-15\cdot3 \pm 0\cdot3$		–	[1]
(methylamino)benzene	298	$-15\cdot7 \pm 0\cdot3$		–	[1]
(dimethylamino)benzene	298	$-15\cdot98 \pm 0\cdot29$		–	[1]
oxole	298·2	$-3\cdot17$		c-C_6H_{12}	[276]
2-methyloxole	298·2	$-3\cdot99$		c-C_6H_{12}	[276]
benzene	298	$-8\cdot2 \pm 1\cdot5$	$-27\cdot2$	c-C_6H_{12}	[94]
	313	$-7\cdot1$		C_7H_{16}	[193]
methylbenzene	313	$-9\cdot2$		C_7H_{16}	[193]
1,3,5-trimethylbenzene	313	$-10\cdot0$		C_7H_{16}	[193]
1-methylnaphthalene	313	$-6\cdot7$		C_7H_{16}	[193]
cyclohexyl isocyanide	294·9	$-8\cdot8$		–	[36]
1,3-dithiolanethione-2	–	$-3\cdot8$		–	[17]
thiolane	296	$-9\cdot5 \pm 3\cdot0$	$-42\cdot7 \pm 2\cdot5$	c-C_6H_{12}	[400]
diethyl sulphide	296	$-7\cdot1 \pm 0\cdot1$	$-36\cdot2 \pm 0\cdot3$	c-C_6H_{12}	[400]
dioctyl sulphide	303·15–333·15	$-8\cdot08$	$-33\cdot8$	–	[352]
dimethyl disulphide	299	$-3\cdot9 \pm 1\cdot1$	-28 ± 4	c-C_6H_{12}	[400]
tetraethylammonium chloride	273	$-39\cdot3$		solid	[103]
tetraethylammonium bromide	273	$-36\cdot0$		solid	[103]
tetrabutylammonium chloride	273	$-58\cdot6$		solid	[103]
	300	$-5\cdot8 \pm 0\cdot2$	$-17\cdot6 \pm 0\cdot4$	CH_3CN	[158]
tetrabutylammonium bromide	273	$-52\cdot7$		solid	[103]
	300	$-3\cdot9 \pm 0\cdot2$	$-15\cdot5 \pm 0\cdot4$	CH_3CN	[158]
tetrabutylammonium iodide	300	$-3\cdot51 \pm 0\cdot04$	$-18\cdot0 \pm 0\cdot4$	CH_3CN	[158]

[a] For 2:1 complex.

TABLE 3.7

Changes in stretching frequency, half-widths, and intensities of the C–H bond of
trichloromethane upon complex formation

Donor	$\dfrac{10^4\,\Delta\nu_1}{\nu_1}$	$\dfrac{w_{1/2}}{cm^{-1}}$	ΔA	Solvent	Ref.
hexane	*49*			C_6H_{14}	[198]
	31(0)			$C_6H_{14}[C_6H_{14}]$	[372]

hexane—*cont.*	*44*			C_6H_{14}	[106]
cyclohexane	*49*			c-C_6H_{12}	[198]
	52			c-C_6H_{12}	[288]
tetrachloromethane	*27(0)*	12	1^b	$CCl_4[HCCl_3]$	[184]
	36			CCl_4	[23]
	38			CCl_4	[198]
	51			CCl_4	[199]
	35(4)	12	2.3^a	$C_6H_{14}[C_6H_{14}]$	[372]
	46			CCl_4	[106]
	52			CCl_4	[288]
	42(0)		1^b	$CCl_4[CCl_4]$	[46]
	44(0)		1^b	$CCl_4[CCl_4]$	[46]
trichloromethane	*27(0)*	11	1.24^b	$DCCl_3[DCCl_3]$	[184]
(self-association)	*55(2)*			$HCCl_3[CCl_4]$	[27]
	86(36)			solid[liquid]	[39]
	49			$DCCl_3$	[198]
	40			$HCCl_3$	[23]
	46			$HCCl_3$	[106]
	31			$DCCl_3$	[106]
	53			$HCCl_3$	[287]
	52			$HCCl_3$	[288]
	94(46)			$HCCl_3[CS_2]$	[78]
	26(4)	12	5.5^a	$C_6H_{14}[C_6H_{14}]$	[372]
	26(0)			$CCl_4[DCCl_3]$	[315]
tribromomethane	*75*			$HCBr_3$	[106]
diiodomethane	*115*			H_2CI_2	[106]
1-chlorobutane	*55*			$C_3H_7CH_2Cl$	[198]
1-bromobutane	*75*			$C_3H_7CH_2Br$	[198]
1-iodobutane	*104*			$C_3H_7CH_2I$	[198]
	102			$C_3H_7CH_2I$	[106]
tetrachloroethene	*43*			$CCl_2=CCl_2$	[23]
chlorobenzene	*64*			C_6H_5Cl	[198]
	64			C_6H_5Cl	[199]
benzene	*66(22)*		5.1^b	$C_6H_6[CCl_4]$	[46]
	49(22)	11	6.2^b	$C_6H_6[HCCl_3]$	[184]
	71			C_6H_6	[198]
	71			C_6H_6	[199]
	53(22)	11	11.7^a	$C_6H_{14}[C_6H_{14}]$	[372]
	64			C_6H_6	[106]
	85			C_6H_6	[288]
methylbenzene	*73*			$C_6H_5CH_3$	[198]
	73			$C_6H_5CH_3$	[199]
1,3-dimethylbenzene	*73*			$C_6H_4(CH_3)_2$	[199]
1,4-dimethylbenzene	*73*			$C_6H_4(CH_3)_2$	[198]
1,3,5-trimethylbenzene	*49(22)*	11	6.8^b	$C_6H_3(CH_3)_3[HCCl_3]$	[184]
	75			$C_6H_3(CH_3)_3$	[198]
	75			$C_6H_3(CH_3)_3$	[199]
1,2,4-trimethylbenzene	*75*			$C_6H_3(CH_3)_3$	[198]
	75			$C_6H_3(CH_3)_3$	[199]
hexamethylbenzene	*86*			$C_6(CH_3)_6$	[199]
methylnitrobenzene	16			$H_3CC_6H_4NO_2$	[288]
nitrobenzene	*44(18)*	13	6.0^b	$C_6H_5NO_2[HCCl_3]$	[184]
nitromethane	*13*			CH_3NO_2	[106]
carbon disulphide	*81*			CS_2	[23]
	86			CS_2	[198]
	80			CS_2	[106]
	86			CS_2	[106]
ethanol	*84(53)*	28	52^a	$C_6H_{14}[C_6H_{14}]$	[372]

As if spoken by a character in a novel—Barthes's line came back to me with special poignancy as I turned from the amateur photographs in my inheritance from Aunt Vandalia, to the studio portraits I had largely ignored when preparing my makeshift lecture on family photography. The wedding pictures of John Mays and Erin Bass, in 1904, spoke eloquently about the world in which the snapshots of children playing, dress-up parties, and such were taken. For those taking part, the photographs seem to say the wedding was more a ritual of caste than a celebration of romance. But it always had been. The earlier representations of marriage in my family—the raw material of genealogy—had been legal documents; and when professional photography came to the South, before the Civil War, the studio artists transposed that bare legal tradition into imagery.

The oldest wedding pictures in Aunt Vandalia's hoard testify to the legal unions of two independent Americans and their respective wealth; they squarely face the studio cameras in clothing black and hard as armor, hands joined with the stiff formality of a handshake. My grandparents' extraordinarily beautiful wedding pictures of 1904 are no less documentary, though the cultural background against which these two young people joined their lives could hardly be more different. There is much richness in lace, and soft light glints off luxuriously draped satin. The faces of the betrothed are relaxed, bright and eager. These sexually blooming, elegantly attired young people have been posed with a rather studied casualness among baronial furnishings, potted tropical plants, romantic paintings, and other props connoting wealth and privilege.

The more I looked at the wedding shots, the more John and Erin appeared to recede into an inventory of props and costumes, becoming props themselves—elements in a visual tableau meant to place their union squarely in the Modern consumer culture their fathers had helped inaugurate in east Texas. There is nothing at all identifiably Southern, by the way, about these pictures. They could have been taken anywhere in the heavy golden haze of the international Edwardian era—in Newport or New Orleans, or in many other American towns no bigger than Henderson. What counted in this new culture was the commodity; and commodities know no nationality, loyalties to place, no fixities.

When John and Erin married, *two* sets of official pictures were made. Only in one did the couple appear. In the other set of large portraits, all people, even the couple, have been banished, leaving products to epitomize the status of the families involved. Each room in the grand Bass house at Palestine was photographed heavy with wedding gifts, arranged in much the same way we might expect to find them in the windows of Captain Mays's Henderson emporium. But only in the light of the popular American beliefs about love and marriage in the late twentieth century do these pictures look strange. The houses of M. B. Bass and Captain Mays were indeed members and full partners in this Gilded Age union. It was only right that the architectural partners should be given the same dignified, professional portrayal as the human ones.

For a long time, of course, I was absent from the pageant. And because photography seemed to cease when my father died, I appear only as a child and little boy. I can hardly recognize myself. The baby dandled on my mother's lap under banana leaves, the boy in a cowboy suit on the porch, or in hiking shorts for an afternoon walk, or put into a panorama of relations for a Christmastide group portrait—none of these portrayals was the person I thought myself to have been, yet *all were*. Like all my family pictures, these had been made by excluding much that now came back to stalk my thoughts—the casual cruelty of the father who happily boosts me up on top of a cotton bale in one photo, the bedeviling hatred of my young life in Shreveport that made me long for escape to that magical place called Texas, where I could dress up like a cowboy all the time, and be free. In the picture, however, I am merely another little boy in a cowboy suit—masked, the sum of costumes and props and backdrops, merely another rhetorical figure in the family story.

But if factual, which I'd already decided they were, did these photographs contain anything *true*, anything revelatory? Had I been asked that question a decade before I found and studied Aunt Vandalia's wealth of photographs, I would surely have said no. For by that time, I had drifted into agreement with the notion that photography does not tell the truth, and reveals nothing—surely the most venerable cliché in the discussion of this art. What I learned from family pictures is that photography tells *nothing* but the truth—but truth that is always contaminated by wishful thinking, sentiment, or dread. It is devious, mischievous truth, as compelling and forever

TABLE 3.7—*continued*

Donor	$\dfrac{10^4\,\Delta\nu_1}{\nu_1}$	$\dfrac{w_{1/2}}{cm^{-1}}$	ΔA	Solvent	Ref.
ethanol—*cont.*	198(149)	13	0.90[c]	$C_2H_5OH[HCCl_3]$	[326]
phenyl methyl ether	60			$C_6H_5OCH_3$	[198]
	61(13)	12	7.5	$C_6H_5OCH_3[DCCl_3]$	[246]
diethyl ether	93(44)	16	11.5	$(C_2H_5)_2O[(C_2H_5)_2O]$	[246]
	96(46)			$(C_2H_5)_2O[(C_2H_5)_2O]$	[78]
	89			$(C_2H_5)_2O$	[198]
	86(46)			$(C_2H_5)_2O[(C_2H_5)_2O]$	[230]
	65	16.2		$(C_2H_5)_2O$	[296]
	80			$(C_2H_5)_2O$	[287]
	71(40)	13	48[a]	$C_6H_{14}[C_6H_{14}]$	[372]
	87			$(C_2H_5)_2O$	[106]
	80			$(C_2H_5)_2O$	[288]
	188(139)	23	1.10[c]	$(C_2H_5)_2O[HCCl_3]$	[326]
dipropyl ether	92	15.9		$(C_3H_7)_2O$	[296]
di(2-propyl) ether	113			$[(CH_3)_2CH]_2O$	[106]
	111	16.5		$[(CH_3)_2CH]_2O$	[296]
dibutyl ether	99			$(C_4H_9)_2O$	[198]
	95	15.2		$(C_4H_9)_2O$	[296]
	97(48)	15	11.5	$(C_4H_9)_2O[DCCl_3]$	[246]
	97(53)		20.2[b]	$(C_4H_9)_2O[CCl_4]$	[46]
	95(52)		10.9[b]	$(C_4H_9)_2O[CCl_4]$	[46]
1,4-dioxane	70(26)		16.5[b]	$O(C_2H_4)_2O[CCl_4]$	[46]
	75(33)		11.1[b]	$O(C_2H_4)_2O[CCl_4]$	[46]
	69	16.8		$O(C_2H_4)_2O$	[296]
	49(18)	14	37[a]	$C_6H_{14}[C_6H_{14}]$	[372]
	71(22)	14	18.0	$O(C_2H_4)_2O[DCCl_3]$	[246]
	224(176)	19	1.20[c]	$O(C_2H_4)_2O[HCCl_3]$	[326]
oxolane	86	18.0		C_4H_8O	[296]
1-methyl-4-(2-propyl)- 1,8-epoxycyclohexane (*eucalyptol*)	141			*eucalyptol*	[106]
propanone	27(0)	15	10.8[b]	$CH_3COCH_3[DCCl_3]$	[184]
	49	14.7		CH_3COCH_3	[296]
	53			CH_3COCH_3	[198]
	80(49)	14	42[a]	$C_6H_{14}[C_6H_{14}]$	[372]
	48(0)	11	12.5	$CH_3COCH_3[DCCl_3]$	[246]
	46			CH_3COCH_3	[106]
	46			CD_3COCD_3	[106]
	92(43)	15	1.10[c]	$CD_3COCD_3[HCCl_3]$	[326]
	49(0)			$CD_3COCD_3[CD_3COCD_3]$	[78]
butanone	53(8)		19.1[b]	$C_2H_5COCH_3[CCl_4]$	[46]
	42(0)		10.6[b]	$C_2H_5COCH_3[CCl_4]$	[46]
cyclohexanone	66(22)		13.2[b]	$C_6H_{10}O[CCl_4]$	[46]
	62(19)		11.3[b]	$C_6H_{10}O[CCl_4]$	[46]
heptanone-4	44	16.7		$C_3H_7COC_3H_7$	[296]
2,6-dimethylheptanone-4	40	20.7		$[(CH_3)_2CH_2CH]_2CO$	[296]
nonanone-5	44	19.9		$C_4H_9COC_4H_9$	[296]
methyl ethanoate	19			CH_3COOCH_3	[288]
ethyl ethanoate	35(−13)	13	9.5	$CH_3COOC_2H_5[DCCl_3]$	[246]
pentyl ethanoate	75(58)	22	12[a]	$C_6H_{14}[CH_3COOC_5H_{11}]$	[206]
ethanoic anhydride	17(−31)	21	8.0	$(CH_3CO)_2O[DCCl_3]$	[246]
dimethyl sulphoxide	132(66)			$DCCl_3[CH_3SOCH_3]$	[324]
	137(84)			$CCl_4[CCl_4]$	[6]

dimethyl sulphoxide–*cont.*	142(96)			$CCl_4[CCl_4]$	[7]
	185(136)			$CD_3SOCD_3[HCCl_3]$	[240]
triethoxyphosphine oxide	118		2·63	$(C_2H_5O)_3PO[CCl_4]$	[163]
tributoxyphosphine oxide	*80(44)*			$CCl_4[DCCl_3]$	[315]
butyldibutoxyphosphine oxide	*106(71)*			$CCl_4[DCCl_3]$	[315]
dibutylbutoxyphosphine oxide	*133(98)*			$CCl_4[DCCl_3]$	[315]
tributylphosphine oxide	*168(133)*			$CCl_4[DCCl_3]$	[315]
	326(210)		9·95	$(C_4H_9)_3PO[CCl_4]$	[163]
ethyl ethanamide	*27(0)*	25	34[b]	$CH_3CONHC_2H_5[DCCl_3]$	[184]
ammonia	0			vapour	[162]
	162(103)			$CCl_4[CCl_4]$	[68]
diethylamine	*345(315)*	53	96·7[a]	$C_6H_{14}[C_6H_{14}]$	[372]
trimethylamine	*0*			vapour	[162]
	366			$(CH_3)_3N[vapour]$	[162]
triethylamine	*388(346)*		53[b]	$(C_2H_5)_3N[CCl_4]$	[46]
	421(380)			$(C_2H_5)_3N[CCl_4]$	[46]
	398(372)	42	36[b]	$(C_2H_5)_3N[DCCl_3]$	[184]
	369			$(C_2H_5)_3N$	[198]
	362(328)	51	70[a]	$C_6H_{14}[C_6H_{14}]$	[372]
	380(337)			$CCl_4[CCl_4]$	[19]
	385			$(C_2H_5)_3N$	[106]
	422			$(C_2H_5)_3N$	[106]
	475(367)			$Cl_2C=CCl_2[Cl_2C=CCl_2]$	[146]
ethanenitrile	39(−3)		7·9[b]	$CH_3CN[CCl_4]$	[46]
benzonitrile	42(0)		6·8[b]	$C_6H_5CN[CCl_4]$	[46]
	48(4)		•9·7[b]	$C_6H_5CN[CCl_4]$	[46]
azine	217(175)		21·7[b]	$C_5H_5N[CCl_4]$	[46]
	227(179)			$C_5D_5N[HCCl_3]$	[240]
	198(152)			$CCl_4[CCl_4]$	[7]
	188			C_5H_5N	[198]
	181(124)			$CCl_4[CCl_4]$	[6]
	181(151)	44	70[a]	$C_6H_{14}[C_6H_{14}]$	[372]
	199			C_5H_5N	[106]
	207(159)	41	20·5	$C_5H_5N[DCCl_3]$	[246]
lithium chloride	*323(267)*			$CH_3COCH_3[CH_3COCH_3]$	[6]
lithium bromide	*327(267)*			$CH_3COCH_3[CH_3COCH_3]$	[6]
sodium iodide	*389(333)*			$CH_3COCH_3[CH_3COCH_3]$	[6]
tetrabutylammonium chloride	*429(377)*			$CCl_4[CCl_4]$	[6]
tetrabutylammonium bromide	*393(333)*			$CCl_4[CCl_4]$	[6]
	469(413)			$CH_3COCH_3[CH_3COCH_3]$	[6]
tetrabutylammonium iodide	*459(395)*			$DCCl_3[DCCl_3]$	[324]
tetraheptylammonium iodide	*349(298)*			$CCl_4[CCl_4]$	[6]
	402(347)			$CH_3COCH_3[CH_3COCH_3]$	[6]
tetrabutylammonium (triphenylphosphino)tri-iodocobaltate(II)	*176(111)*			$DCCl_3[DCCl_3]$	[324]
tetrabutylammonium (triphenylphosphino)tri-iodonickelate(II)	*181(115)*			$DCCl_3[DCCl_3]$	[324]
tetrabutylammonium (triphenylphosphino)tri-iodozincate(II)	*176(111)*			$DCCl_3[DCCl_3]$	[324]
octyne-4	*75(31)*			$CCl_4[CCl_4]$	[186]

Frequency shifts in italics indicate values for $DCCl_3$.

[a] $\Delta A = A$ (in hexane + donor)$/A$(in hexane).

[b] $\Delta A = A$ (in solvent)$/A$ (in CCl_4).

[c] $\Delta A = A$ (in solvent)$/A$ (in $HCCl_3$).

TABLE 3.8

Changes in bending frequency, and half-widths for the C–H bond of trichloromethane

Donor	$\dfrac{10^4 \, \Delta \nu_4}{\nu_4}$	$w_{1/2}/cm^{-1}$	Solvent	Ref.
hexane	− 29		C_6H_{14}	[106]
	−54		C_6H_{14}	[106]
	−22	10	C_6H_{14}	[372]
heptane	−40		C_7H_{16}	[46]
	−43		C_7H_{16}	[46]
carbon disulphide	− 53		CS_2	[106]
	−82		CS_2	[106]
tetrachloromethane	*−11*	9·0	C_6H_{14}	[372]
	−49		CCl_4	[106]
	−43		CCl_4	[46]
	−25		CCl_4	[106]
	−40		CCl_4	[46]
trichloromethane	*0*	10·0	C_6H_{14}	[372]
	−43		$DCCl_3$	[106]
	− 29		$HCCl_3$	[106]
tribromomethane	− 65		$HCBr_3$	[106]
	− 33		$HCBr_3$	[106]
diiodomethane	−82		H_2Cl_2	[106]
	−49		H_2Cl_2	[106]
1-iodobutane	− 38		$C_3H_7CH_2I$	[106]
	− 21		$C_3H_7CH_2I$	[106]
benzene	− 38		C_6H_6	[106]
	−43		C_6H_6	[46]
	− 8		C_6H_6	[106]
	− 8		C_6H_6	[46]
nitromethane	*0*		CH_3NO_2	[106]
	25		CH_3NO_2	[106]
ethanol	*143*	16	C_6H_{14}	[372]
diethyl ether	*121*	23	C_6H_{14}	[372]
	170		$(C_2H_5)_2O$	[106]
	164		$(C_2H_5)_2O$	[106]
	165		$(C_2H_5)_2O$	[78]
di(2-propyl) ether	*186*		$[(CH_3)_2CH]_2O$	[106]
	180		$[(CH_3)_2CH]_2O$	[106]
1,4-dioxane	*110*	13	C_6H_{14}	[372]
1-methyl-4-(2-propyl)-1,8-epoxy-cyclohexane (*eucalyptol*)	213		*eucalyptol*	[106]
	186		*eucalyptol*	[106]
propanone	*110*	15	C_6H_{14}	[372]
	115		CH_3COCH_3	[106]
pentyl ethanoate	*99*	17·0	C_6H_{14}	[372]
dimethyl sulphoxide	*219*		CH_3SOCH_3	[106]
	254		CH_3SOCH_3	[106]
ethanenitrile	*109*		CH_3CN	[106]
	98		CH_3CN	[106]
diethylamine	*308*	26	C_6H_{14}	[372]
triethylamine	*319*	23	C_6H_{14}	[372]
	339		$(C_2H_5)_3N$	[106]

Values in italics are for the C–D bond of DCCl₃.

open to interpretation as the figures in poetry and painting, or the ambiguous messages delivered to ancient Greek pilgrims by the Oracle. "The Lord who prophesies at Delphoi neither speaks clearly nor hides his meaning completely," says Heraclitus. "He gives one symbols instead."

Of all the ancestors I would meet in the terrain of their four hundred years in America, those in east Texas were the first to live entirely within the same world that I do, defined by mass production and consumerism—what Bourdieu calls "the age of frivolity," the gift of advanced capitalism to the generations who came of age after 1890.

Unlike the Squire's children in Mayesville, reared to perpetuate a conventional Southern world of stable agrarian values and Classical ideals of the good life, the sons and daughters of Captain Mays were born within this cosmopolitan Modern culture, writ small on the hills of east Texas. They came into the world at Henderson as citizens of Modernity, heirs to the late nineteenth century's newly aesthetic culture of luxury, excess, and style—if not victims of the famous anxiety that afflicted the sharpest thinkers in bourgeois Europe during its *fin de siècle*. They dressed and groomed and entertained themselves not much differently than did similarly well-off, unthinking young people in Berlin or Boston, St. Petersburg or San Antonio. Like their contemporaries in, say, London or Chicago, they sat through the same saccharine songs, performed by more or less equally untalented cousins at identical after-dinner musicales, all in the name of a propriety believed to be necessary to their station in society. And they all did so in parlors ornamented by the bric-a-brac, potted palms, and such ordained by middle-brow Victorian culture— a persuasive phenomenon that, by the last quarter of the nineteenth century, pervaded the popular imagination of all the world's ascendant commercial empires, both European and American.

It was all new—incredibly new to men born, like my great-grandfather, in the rural South before the Civil War. One sees the impact of the new world reflected in the uncomprehending, strangely drifting expressions of the oldest forebears in the photographs. None seems quite aware that cameras require the same novel things all consumerism requires: posing, alert gazing, attention to one's sensuous appearance to the eye of the other. Their grandchildren, such as

John Matthew Mays and Erin Bass, his bride, were alive to the new passions quickened by what they saw in shop windows and on the shelves of my merchant-grandfather's stores; you can see it in their happy eyes.

Henderson surely lay far from the New South's burgeoning metropolitan centers, but the family home of Captain Mays, its Ionic columns gracefully tall, its front verandah appropriately large, was a busy intersection of books and popular magazines, cameras, and mass-produced musical instruments, stylish clothes and fashionable toys, and numerous other delights from international high consumerism's Gilded Age in America. And as my grandfather and aunts admitted these delightful technologies to the center of their lives, and passed them on to their children—to my father and Vandalia and the rest—they fabricated the culture of youthful happiness suffusing my family archive of photographs.

When, around 1910, John and Sister Erin moved with their two small children from Texas into the traditional Southern ethos of Greenwood, they became Southerners, for that was the way to get on. Though neither had much direct experience of the rural South, they recalled the old memories, codes, and customs passed down by the ancestors—the scripts required of a gentleman and matron in a conservative Southern village at the center of a quilt of farms and plantations—and they played their leading roles in the pageant of Greenwood's life with charm and aplomb, and to high success. My grandfather farmed, as Southern men aspiring to political prominence were supposed to, but left the actual work of farming to others. His natural place was in the village and its society, in the store he ran in Greenwood, and among his business partners in oil and cotton ginning.

Cameras were not, of course, the only festive technologies young John and Vannie enjoyed in their newfound home. While undoing my aunt's house, we found parts of the attic stacked almost rafter-high with the consumer magazines and illustrated mail-order catalogues, advertising the latest hats and dresses, lotions and other beauty aids, electromagnetic weight-reducing belts, undergarments said to be Parisian, sensational time-saving household appliances, fast cars with finely hatted ladies at the steering wheels. All these commodities came trundling into Greenwood on the railway and

triethylamine—*cont.*

	263	$(C_2H_5)_3N$	[46]
	410	$(C_2H_5)_3N$	[106]
	327	$(C_2H_5)_3N$	[46]
azine	*154*	27	C_6H_{14}	[372]
	143		C_5H_5N	[106]

Values in italics are for the C–D bond of $DCCl_3$.

TABLE 3.9

Isotope and solvent effects on the vibrational mode ν_2 of $HCCl_3$

Solvent	ν_2/cm^{-1}			
	$HCCl_3$	Ref.	$DCCl_3$	Ref.
vapour	674·5	[106]	655·5	[106]
	678	[194]	655	[216]
hexane	671	[106]	652·5	[106]
	671	[194]	650·5	[290]
cyclohexane	670·5	[106]	652	[106]
	670·5	[290]	650·5	[290]
carbon disulphide	669	[106]	650·5	[106]
	669	[290]	649·5	[290]
1-iodobutane	668·5	[106]	649·5	[106]
trichloromethane	669·5	[106]	651	[106]
tetrachloromethane	670	[106]	652	[106]
	669·5	[290]	650	[290]
tetrachloroethene	669·5	[290]	650	[290]
diethyl ether	667	[106]	647·5	[106]
	665	[290]	645·5	[290]
di(2-propyl) ether	666·5	[106]	647	[106]
1,4-dioxane	665	[290]	646	[290]
1-methyl-4-(2-propyl)-1,8-epoxycyclohexane				
(*eucalyptol*)	665·5	[106]	645	[106]
ethyl ethanoate	665·5	[290]	646·5	[290]
propanone	668	[106]	648	[106]
	665·5	[290]	646·5	[290]
dimethyl sulphoxide	668	[106]		
ethanenitrile	668·5	[106]	649·5	[106]
	666	[290]	647·5	[290]
triethylamine	661	[106]	641·5	[106]
	659	[290]	640	[290]
tetrabutylammonium bromide in				
tetrachloromethane	659·5	[106]	640·5	[106]

TABLE 3.10

Changes in the carbonyl stretching frequency of donor molecules in complexes with trichloromethane

Donor	$\dfrac{\nu_s(C=O)}{cm^{-1}}$	$\dfrac{10^4 \, \Delta\nu_s}{\nu_s}$ 1:1	$\dfrac{10^4 \, \Delta\nu_s}{\nu_s}$ 1:2	Solvent	Ref.
propanone	1722·0	26	66	c-C_6H_{12}	[396]
cyclohexanone	1724	41	104	c-C_6H_{12}	[83]
	1723·6	42	107	c-C_6H_{12}	[396]

into Sister Erin's house in the teens and twenties of our century; and a sleek, fast car always stood in the driveway.

In the 1920s, when my father, at age sixteen, bought his tiny first airplane for $1,000 and scandalized old Greenwood society by flying dangerously close to the housetops, both John and Aunt Vandalia fell in love with Hollywood. Into one of her "memory books"— albums of clippings, programs for concerts and baseball games, invitations to galas and dress-up banquets, souvenir cigarettes, and such—she lovingly pasted a notice for *The Sheik* (with the handwritten note: "Saw it twice!"), and others for Mary Pickford's *Little Lord Fauntleroy* and De Mille's *Nice People*. (Exclaimed the ad: "Is this mad age? Are we all running Wild? A dramatic expose of the jazz age of today!")

To the gala whirl of cinema and beauty products, aircraft and cars in which my father and aunt came of age, one ingredient was added unceasingly, at Sister Erin's insistence. It was music. My father studied violin, my aunt voice and piano—apparently without a trace of that reluctance we too often find in our children now. To be musical then, after all, was to be Modern; and the children of John and Erin wanted to be very Modern. Music, as much as shopping, church, and socializing, dominated the family calendar. The family rose early one cold morning in January, 1923, to make the train trip into Shreveport to hear Rachmaninoff perform Chopin, Liszt, and his own compositions. They went into town to hear Jascha Heifetz play Mozart and Beethoven, and other stars who regularly passed through the riverside metropolis. At home, the performances were less elevated. Aunt Vandalia's after-dinner songs included "Just Awearyin' for You" (from *Seven Songs As Unpretentious As the Wild Rose*) by Carrie Jacobs-Bond, and Charles Wakefield Cadman's "At Dawning I Love You"; and Sister Erin was famous in the neighborhood for her piano renditions of *The Ride of the Buffaloes* and L. M. Gottschalk's *The Last Hope*. Apart from piano and violin, the favorite musical technology was the wind-up Victrola, from the scratchy throat of which sang Caruso, along with myriad songsters and opera stars now long forgotten.

Then John and Vandalia grew up, and married, both in the late 1920s. My father and his bride moved to Spring Ridge, Vannie and Alvin into the cottage close to the house in Greenwood. Then came the Crash.

And with the coming of the Great Depression, which ruined the cotton economy in which my grandfather had prospered, Vandalia pasted the last memento of happy days into her memory books, and shut them forever. The heavy Victrola records were no longer played, no longer bought, the deluxe mail-order catalogues were dispatched to the attic, and the house became still. Like everyone who had done well in Southern villages throughout the 1920s, my grandparents suffered from the precipitous slide in prices and the Depression-era shortages, but were hardly pauperized by them. The careful, patient founding of their Greenwood reputations and circle of friendships sustained them through the early, dark years. My grandfather regained his footing in politics and business, and Sister Erin reclaimed hers in the village's Methodist matriarchy. But perhaps because they had known no other life than one of Modern wealth and freedom, and enjoyed this comfortable, pleasurable life in Greenwood from their earliest years, John and Sister Erin's children appear to have emerged from the first shocks of the Depression peculiarly damaged, shamed, bewildered by the changes. As the family's fortunes recovered, the spirits of my father and aunt did not. These children of the Jazz Age drifted through the next decade in an attenuated daze, going through the motions of living, but without joy, interest, enthusiasm.

From my earliest years, that is how I remember beloved Aunt Vandalia; and that is how she remained until the social tumults of the 1960s focused her mind on genealogy, and the illnesses of Sister Erin and Uncle Alvin on cures and God.

As for my father, I do not know whether he ever found moorings, before he drifted away from me into final night and silence.

I thought I knew him. But what I imagined of my father was hallucination, concocted from Sister Erin's adulatory reminiscing, tales in the hero books he had loved as a boy, the complimentary things his surviving friends and relatives had to say about him. The intense, first years of psychotherapy helped loosen the malignant barnacles on the hull of my mind, sweeping away the toxic fantasies I had so long clutched close and leaving me with only scattered images of my father, and no continuous knowledge of him at all.

An unsorted box of mental pictures, like snapshots, is what now

TABLE 3.10–*continued*

Donor	$\nu_s(C=O)$ $\overline{cm^{-1}}$	$\dfrac{10^4\,\Delta\nu_s}{\nu_s}$ 1:1	$\dfrac{10^4\,\Delta\nu_s}{\nu_s}$ 1:2	Solvent	Ref.
heptanone-4	1721	52	90	C_6H_{14}	[83]
bornanone	1753	57	103	C_6H_{14}	[83]
2,2,4,4-tetramethylpentanone-3	1691	59	106	C_6H_{14}	[83]
butyl 2-methylpropanoate	1741	69	144	C_6H_{14}	[83]
ethyl ethanoate	1749·7	61	104	HCCl₃	[205]
dimethyl ethanamide	1674	78	185	C_6H_{14}	[83]
methyl ethanamide	1687	119		HCCl₃	[218]
1-methylazolidinone-2	1714	88	181	C_6H_{14}	[83]
1,3-diethyl 1,3-diphenyl urea	1666	108	204	C_6H_{14}	[83]
1,3-dimethyl urea	1668	102	204	C_6H_{14}	[83]
'propionopiperidide'	1667	102	210	C_6H_{14}	[83]
'isobutyropiperidide'	1661	102	217	C_6H_{14}	[83]

TABLE 3.11

Solvent effects on the ^{1}H nuclear magnetic resonance of trichloromethane

Initial medium	Chemical shift difference/ppm	Final medium	Ref.
cyclohexane	0·01	petroleum ether	[367]
hexane	0·00	heptane	[136]
hexane	0·03	octane	[136]
hexane	0·01	2-methylheptane	[136]
hexane	0·11	cyclohexane	[136]
hexane	0·26	bicyclo[4.4.0]decane	[136]
'paraffins'	0·22	'chlorinated hydrocarbons'	[135]
cyclohexane	0·120	tetrachloromethane	[242]
cyclohexane	0·17	tetrachloromethane	[263]
hexane	0·48	tetrachloromethane	[136]
vapour	0·45₂	tetrachloromethane	[392]
tetrachloromethane	*1·8 ± 0·3*	tetrachloromethane *in trichloromethane*	[180], [200]
cyclopentane	0·29	trichloromethane	[207]
hexane	0·29	trichloromethane	[207]
cyclohexane	0·29	trichloromethane	[207]
cyclohexane	0·20	trichloromethane	[257]
cyclohexane	0·19	trichloromethane	[200]
cyclohexane	0·17	trichloromethane	[182]

Values in italics indicate 'complex shifts'; i.e. the difference in the shift of trichloromethane in the complex from that in the 'free' molecule.

[a] T = 293 K.
[b] T = 308 K.
[c] T = 323 K.
[d] T = 254 K.
[e] T = 272 K.
[f] T = 289 K.
[g] T = 309 K.

cyclohexane	0·147	trichloromethane	[242]
hexane	0·55	trichloromethane	[136]
cyclohexane	0·17	trichloromethane	[263]
cyclohexane	*0·93 ± 0·08*	trichloromethane *in cyclohexane*	[400]
cyclohexane	*1·55 ± 0·5*	trichloromethane *in cyclohexane*	[200]
cyclohexane	*1·8 ± 0·3*	trichloromethane *in cyclohexane*	[180]
tetrachloromethane	0·29	trichloromethane	[207]
tetrachloromethane	0·20	trichloromethane	[136]
tetrachloromethane	0·022	trichloromethane	[200]
cyclohexane	0·06	trichlorosilane	[182]
cyclohexane	*0·44*	dichloromethane *in cyclohexane*	[291]
hexane	0·06	dichloromethane	[136]
vapour	0·44$_3$	dichloromethane	[392]
vapour	0·61$_2$	dibromomethane	[392]
vapour	0·83$_8$	diiodomethane	[392]
hexane	0·76	1,1,2,2-tetrachloroethane	[136]
hexane	0·63	1,2-dichloroethane	[136]
hexane	0·33	chloroethane	[136]
hexane	0·36	1-chlorobutane	[136]
hexane	0·48	chlorocyclohexane	[136]
vapour	0·67$_7$	iodomethane	[392]
vapour	0·54$_3$	carbon disulphide	[392]
'hydrocarbons'	0·33	1-fluoropropane	[337]
cyclohexane	0·17	hexene-1	[323]
cyclohexane	0·17	cyclohexene	[323]
cyclohexane	0·20	cyclohexadiene-1,3	[323]
cyclohexane	− 0·747	benzene	[242]
cyclohexane	− 0·733	benzene	[367]
cyclohexane	− 1·35	benzene	[182], [323]
trichloromethane	− 1·07	benzene	[283]
cyclohexane	*− 1·91 ± 0·40*	benzene *in cyclohexane*	[94]
heptane	*− 1·49*	benzene *in heptane*	[193]
tetradecane	*− 1·38$_8$*	benzene *in tetradecane*	[178]
hexadecane	*− 1·39$_7$*	benzene *in hexadecane*	[178]
cis-bicyclo[4.4.0]decane	*− 1·39$_0$*	benzene *in cis-bicyclo[4.4.0]decane*	[178]
1,1'-bicyclohexane	*− 1·33$_8$*	benzene *in 1,1'-bicyclohexane*	[178]
cyclohexane	*− 1·49$_5$*	benzene *in cyclohexane*	[178]
cyclohexane	− 1·35	methylbenzene	[323]
heptane	*− 1·53*	methylbenzene *in heptane*	[193]
cyclohexane	*− 1·59$_8$*	1,4-dimethylbenzene *in cyclohexane*	[178]
cyclohexane	*− 1·78$_8$*	1,4-diethylbenzene *in cyclohexane*	[178]
cyclohexane	− 1·43	1,3,5-trimethylbenzene	[323]
trichloromethane	− 1·4	1,3,5-trimethylbenzene	[337]

TABLE 3.11–*continued*

Initial medium	Chemical shift difference/ppm	Final medium	Ref.
heptane	-1.77	1,3,5-trimethylbenzene *in heptane*	[193]
cyclohexane	-2.05_3	1,3,5-trimethylbenzene *in cyclohexane*	[178]
cyclohexane	-1.68_7	1,3,5-triethylbenzene *in cyclohexane*	[178]
cyclohexane	-1.55_2	1,3,5-tri(2-propyl)benzene *in cyclohexane*	[178]
heptane	-1.95	1-methylnaphthalene *in heptane*	[193]
cyclohexane	-0.01	nitrobenzene	[323]
cyclohexane	-0.39	bromobenzene	[323]
cyclohexane	-0.74	chlorobenzene	[323]
cyclohexane	-0.41	1,2-dichlorobenzene	[323]
trichloromethane	-0.4	1,2-dichlorobenzene	[337]
cyclohexane	-0.05_8	4-fluoro-1-aminobenzene	[367]
cyclohexane	-0.15_0	3-chloro-1-aminobenzene	[367]
cyclohexane	-0.23 ± 0.03	2-chloro-1-aminobenzene	[367]
cyclohexane	-0.14_2	3-methoxy-1-aminobenzene	[367]
cyclohexane	-0.21_7	2-methoxy-1-aminobenzene	[367]
cyclohexane	-0.52 ± 0.03	3-methyl-1-aminobenzene	[367]
cyclohexane	-0.54 ± 0.03	2-methyl-1-aminobenzene	[367]
cyclohexane	-0.52 ± 0.03	2-ethyl-1-aminobenzene	[367]
cyclohexane	-0.80_0	2,4,6-trimethyl-1-aminobenzene	[367]
cyclohexane	-0.15_0	4-chloro-1-(methylamino)benzene	[367]
cyclohexane	-0.50_0	(methylamino)benzene	[367]
cyclohexane	-0.65_0	diphenylamine [*melt*]	[367]
cyclohexane	-0.75_8	(diethylamino)benzene	[367]
cyclohexane	-0.80_0	(dimethylamino)benzene	[367]
trichloromethane	-0.97	(dimethylamino)benzene	[26]
cyclohexane	-0.35 ± 0.03	aminobenzene	[367]
trichloromethane	-0.62	aminobenzene	[26]
'paraffins'	0.53	'nitriles'	[135]
trichloromethane	0.77_7	cyclohexanecarbonitrile	[37]
trichloromethane	0.80	cyclohexanecarbonitrile	[36]
vapour	0.77_7	ethanenitrile	[392]
hexane	0.46	ethanenitrile	[136]
cyclohexane	0.445	ethanenitrile	[242]
trichloromethane	0.53	ethanenitrile	[136]
trichloromethane	0.63	ethanenitrile	[35]
tetrachloromethane	0.973 ± 0.019	ethanenitrile	[180]
hexane	0.48	propanenitrile	[136]
trichloromethane	0.53	propanenitrile	[136]
trichloromethane	0.55	propanenitrile	[337]
hexane	0.53	butanenitrile	[136]
hexane	0.48	2,2-dimethylpropanenitrile	[136]
hexane	0.60	hexanenitrile	[136]
cyclohexane	1.282	azine	[242]
cyclohexane	1.36_7	azine	[367]
trichloromethane	3.90	azine	[35]

remains: my father looming over me, smiling, dressed in the khakis he always wore during the summer growing season at Spring Ridge; my father handing me a glass of orange juice before disappearing into the Shreveport motel suite's other bedroom with a woman he had picked up in a bar; my father beside me, his arms around me, as I fall asleep in the big bed in Spring Ridge. What remains in fact—the only truth about him outside my mind—is the handful of family pictures I found in the attic, and an album of photographs from his hunting expeditions.

I cannot recall how the hunting album came into my keeping, or when. It happened long before my aunt's death, a gift perhaps from Sister Erin in some troubled passage of my life I now find hard to remember much about. When we cleaned up Aunt Vandalia's house, it was joined by myriad other photos in which my father appears. But of all the images of him I own, none are more tightly ringed by that magic fire of anxiety and kinship that burns in the interval between fathers and sons.

As I write these words, the album lies on my desk, open to the first page. The bold handwritten inscription inside the front cover, inscribed in white ink on black paper, reads: *To Dad from John Bass.* I suspect my father had no intention of putting together his photographs into a scrapbook when the trips began in 1928. He had always loved taking pictures with his fine box camera, and anyone in the family would have found it odd had he not carried it along. Only after the last trip he documented, in 1936, did he lift the photos from the chaos of drawer or shoebox, sort and mount and label them, and give them to my grandfather.

By that time, he had shot many rolls of film in the desert mountains and wildernesses of west Texas and northern Mexico, and on the peninsula of Baja California, during the winter hunts in the years between 1928 to 1936. (For a reason unknown to me—a year without a hunt? a lost or damaged camera, a ruined roll of film?—there are no photos from 1935.) The 1936 trip was not the last one. I know my father's expeditions went on past 1941, the year of my birth, for I remember his returns, and the gifts of fantastically carved, painted walking sticks and toys he fetched home for me from Mexico.

The first voyage on which John Bass took along his camera was in the year after his marriage to my mother. Late autumn, 1928: John Bass and his father, and their male relatives and friends and servants, set out by car over rough, narrow roads through the Louisiana hills toward the dry mesquite barrens of Texas, and beyond. For a month, the men hunted. The next year, they rambled farther south and west. By 1931, they were ranging south over the border into Mexico, following the lure of deer, quail, coyote. By 1936 they had several times pushed their car across the desert mountains of Mexico's Chihuahua region and into Sonora, where they hunted mountain sheep, eagles, and dry-range deer, then fished in the Gulf of California and off the Pacific coast. During the nearer trips, they drove to the end of the roads, and camped. On the farther voyages, later in the 1930s, John Bass and the men based themselves at seasonal hunting camps—*Mr. Ren's, the Jap's,* in my father notation—isolated adobe huts in the wilderness, it seems, where they were provisioned and given guides for the hunts beyond the last roads.

The album has no title leaf—only a picture of a dry patch of prairie littered with dead birds, over which a late afternoon shadow of my father's tall, slender form falls, like a signature. On the first page he wrote *Cotulla, Texas 1928* above photos of animal corpses. In two of them, a bobcat and a white-tail deer—*a nice buck* reads the caption—appear, the cat strung up on a barbed-wire fence, the buck attached to a dry tree by nails driven through its muzzle. In another, a smiling black servant labeled Bennie, mugging for the camera, holds up the flaccid pelted body of another bobcat. Throughout the book are brief legends:

> *First nite. 30 miles south of Columbus, N.M. Ice in the water bucket. New sleeping bags OK.*
> *J. B. holding a California quail. We killed all we wanted.*
> *Where the ram fell.*
> *Crippled Coyote.*
> *J. B. and a coyote he killed at 190 yds.*
> *J. B. and coyote.*
> *J. M. and crippled coyote.*
> *The cripple.*
> *J. M., Am & crippled coyote.*

Many other names come into the captions, some known to me. "J. M." is John Matthew Mays, my grandfather. "Willis"—a magnifi-

trichloromethane	*3·73*	azine	[36]
cyclohexane/tetrachloromethane	*2·089*	azine *in cyclohexane/tetrachloromethane*	[125]
tetrachloromethane	*1·07 ± 0·19*	azine *in tetrachloromethane*	[158]
tetrachloromethane	*2·271 ± 0·023*	azine *in tetrachloromethane*	[180]
cyclohexane	1·30 ± 0·03	2-methylazine	[367]
cyclohexane	$1·56_7$	2,6-dimethylazine	[367]
cyclohexane	$1·66_7$	2,4,6-trimethylazine	[367]
trichloromethane	1·47	cyclohexylamine	[283]
hexane	1·63	methylamine	[136]
trichloromethane	2·20	methylamine	[58]
hexane	1·63	propylamine	[136]
trichloromethane	2·30	propylamine	[58]
cyclohexane	$1·63_3$	propylamine	[367]
hexane	1·63	2-propylamine	[136]
hexane	1·60	butylamine	[136]
cyclohexane	$1·56_7$	butylamine	[367]
trichloromethane	2·31	butylamine	[58]
cyclohexane	$1·63_3$	2-methyl-2-propylamine	[367]
hexane	1·45	2-methyl-2-propylamine	[134]
hexane	1·56	2-methyl-2-propylamine	[136]
cyclohexane	$0·76_7$	benzylamine	[367]
cyclohexane	$1·40_0$	3-methoxypropylamine	[367]
cyclohexane	$1·36_7$	2-methoxyethylamine	[367]
cyclohexane	$1·35_0$	2-propenylamine	[367]
cyclohexane	$1·25_8$	(diethoxymethyl)amine	[367]
cyclohexane	$1·22_5$	2-hydroxypropylamine	[367]
cyclohexane	$1·21_7$	2-hydroxyethylamine	[367]
cyclohexane	$0·99_2$	2-(aminomethyl)oxole	[367]
cyclohexane	$0·81_7$	3-aminopropanenitrile	[367]
cyclohexane	$0·41_7$	aminomethanenitrile	[367]
cyclohexane	$0·15_8$	1,2-diphenylethylamine	[367]
cyclohexane	$-0·18_3$	O-benzylhydroxylamine	[367]
cyclohexane	$-0·31_7$	dibenzylamine	[367]
trichloromethane	2·51	dimethylamine	[58]
cyclohexane	$1·53_3$	diethylamine	[367]
hexane	1·48	diethylamine	[134]
hexane	1·51	diethylamine	[136]
hexane	1·25	dipropylamine	[134]
cyclohexane	$1·29_5$	di(2-propyl)amine	[367]
hexane	1·27	di(2-propyl)amine	[367]
hexane	1·35	di(2-propyl)amine	[136]
hexane	1·35	dibutylamine	[136]
hexane	1·22	dibutylamine	[134]
hexane	0·93	di(2-methylpropyl)amine	[136]
hexane	0·83	di(2-methylpropyl)amine	[134]
hexane	1·07	dicyclohexylamine	[134]
cyclohexane	$1·29_5$	methyl-2-propenylamine	[367]
cyclohexane	$1·21_7$	tetrahydro-1,4-oxazine	[367]
cyclohexane	$0·95_8$	di(2-propenyl)amine	[367]
cyclohexane	$1·15_8$	di(2-methoxyethyl)amine	[367]
cyclohexane	$1·15_0$	methyl-2,2-diethoxyethylamine	[367]
cyclohexane	$1·03_3$	(methylaminomethyl)oxole	[367]
cyclohexane	0·93 ± 0·05	di(2-hydroxyethyl)amine	[367]

cent man riding bareback, pulling up his horse just short of the camera—is Willis Hill, the white overseer of my father's plantation. "Pat" may be Pat Camp, a cousin and contemporary of my father, who practiced law in San Antonio. And there is "Old Am," a black servant who regularly accompanied the expeditions—*Old Man Am*, my father once calls him. But who was Mr. Ren, who frequently appears in the album, holding up a dead eagle for the camera, posing beside a felled buck?

When I was an adolescent, and found out about such matters—I had the album by then—I imagined the broad, low stucco building in the photograph captioned *Mr Rens club* to be populated by voluptuous Mexican whores, available to the hunters when they tired of killing. Apart from erotic fantasies, then and later, the pictures prompted a persisting notion that Mr. Ren or the Jap or other guides and suppliers in the photos might still be alive, very old but holding memories I could harvest and keep. I thought that I could study the topography in each photo until I had it memorized, then travel in the western American wastelands until I discovered those men and that landscape, and extract from them memories of my dead father. If I failed to find the storytellers, I could at least find the spots where the stories came from, the rocky expanses of desert emptiness that my father surely loved more than any other places on earth—more, surely, than Spring Ridge, which he often left.

For if born a Texan, my father was reared a Southerner, and brought up by Sister Erin, herself a stranger to such a life, to be a squire and farmer of the old Southern sort—an extension into a new generation, that is, of the virtuous old ideas she had inherited, then translated into her new life at Greenwood. He learned his lessons from his mother with perhaps too much outward deference, for an opposite urge stirred within him. His early reading had included *Sports and Pastimes—Indoors and Out*, one volume in the *Our Wonder World* series, which I brought back home with me after Vandalia died. The following definition of male youth from its pages appears to have defined his own self-understanding: "Could we but get a cross-section of a normal boy's mental make-up, we would find there a little barbarism, some savagery, a little knighthood, here and there a willingness to do what others like, much hatred for convention and sham, some desire to be conspicuous, some greed, some respectful tolerance, and a superabundance of love for adventure and fun." On his later western voyages, John Bass liked to be photographed in the pose of such a young Noble Savage, as he'd learned this iconography from *Our Wonder World* and the other illustrated adventure books he had devoured as a child: Man in the stately, relaxed attitude of natural mastery, holding his gun like a scepter, towering over the prostrate corpse of some beast he had killed.

Such men need heraldry, and opportunities for chivalry beyond hunting down wild animals. Dispatched by Sister Erin to a military college for a year or two, my father was afterward discouraged from taking up the military career he was probably best suited for; for he was good with guns, liked the feel and heft and ownership of them, and he was a good pilot. Instead, Sister Erin saw to it that their son was given Spring Ridge as a wedding present, and sent there to become a cotton planter and merchant, and a rural gentleman of the sort no Mays ancestor had been since before the Civil War.

Everything about the scenario was wrong. My father entered the world and came of age in the Modern age, defined by the technologies of pleasure and mass production, only to find himself, still a young man, in a dying caste of Southern cotton planters, doomed to extinction. And the woman John Bass married, the mother of his three children, was a vivid, beautiful Christian Scientist, the daughter of a Shreveport family enriched by the city's real-estate industry—less suited for genteel farm life, that is, than even my father.

I knew my mother little better than I knew John Bass; and most of what I do recall is shot through with pain and silence, for it was learned during the long illness that took her life when I was nine years old. But it was never a secret, even among my Greenwood relatives, that my father abandoned his wife, his children, and his plantation with casual cruelty at every opportunity, for periods long and short. He always had business elsewhere, or visiting to do, or Democratic Party functions to attend (for, like his father, he was a loyal New Deal Democrat). He was refused admission to the air force at the outbreak of the Second World War because of his age and occupation as a farmer. But he was able to take part, in a sense, by becoming an informal agent of the Federal Bureau of Investigation, charged with looking out for signs of Nazi subversion among the rural blacks—busywork, of course, but yet another pretext, of the

TABLE 3.11–*continued*

Initial medium	Chemical shift difference/ppm	Final medium	Ref.
cyclohexane	0.85_0	methylamino-β-alanylnitrile	[367]
cyclohexane	0.73_3	ethyl methylaminocrotonate	[367]
cyclohexane	0.63_3	methylaminoethanenitrile	[367]
cyclohexane	0.60_0	di(2-propynyl)amine	[367]
cyclohexane	0.53_3	di(2-cyanoethyl)amine	[367]
cyclohexane	0.71_7	methylbenzylamine	[367]
cyclohexane	0.31_7	benzyl(2-hydroxyethyl)amine	[367]
hexane	1.25	trimethylamine	[136]
hexane	1.23	trimethylamine	[134]
hexane	1.24	triethylamine	[136]
hexane	1.08	triethylamine	[134]
cyclohexane	0.967	triethylamine	[242]
hydrocarbon	1.20	triethylamine	[337]
tetrachloromethane	1.25	triethylamine	[136]
trichloromethane	0.28	triethylamine	[207]
cyclohexane	1.48 ± 0.04	triethylamine *in cyclohexane*	[94]
cyclohexane	1.472 ± 0.011	triethylamine *in cyclohexane*	[180]
cyclohexane	1.592 ± 0.007	triethylamine *in cyclohexane*	[400]
hexane	0.60	tripropylamine	[134]
hexane	0.78	tributylamine	[136]
hexane	0.63	tributylamine	[134]
cyclohexane	*1.60*	tributylamine	[289]
hexane	0.66	ethyldi(2-propyl)amine	[136]
hexane	0.35	butyldi(2-methylpropyl)amine	[136]
hexane	0.08_3	butyldi(2-methylpropyl)amine	[134]
cyclohexane	*1.67*	trihexylamine *in cyclohexane*	[289]
cyclohexane	0.60_0	tris(2-propenyl)amine	[367]
cyclohexane	*1.90*	trioctylamine *in cyclohexane*	[289]
hexane	0.83	cyclohexyldiethylamine	[134]
trichloromethane	1.11	cyclohexyldimethylamine	[283]
hexane	0.40	cyclohexyldibutylamine	[134]
trichloromethane	*2.56*	perhydroazepine	[58]
trichloromethane	*2.58*	perhydroazine	[58]
cyclohexane	1.63_3	perhydroazine	[367]
hexane	1.73	perhydroazine	[136]
hexane	1.48	2,2,4,4-tetramethylperhydroazine	[136]
trichloromethane	*2.68*	azolidine	[58]
trichloromethane	2.05	1-methylazolidine	[37]
trichloromethane	2.05	1-methylazolidine	[35]
trichloromethane	2.13	1-methylazolidine	[36]
trichloromethane	*3.05*	azetidine	[58]
trichloromethane	*2.24*	aziridine	[58]
cyclohexane	-0.75_0	azole	[367]
cyclohexane	-0.94_2	indole [*melt*]	[367]

trichloromethane	*1·98*	N-ethylidene-2-propylamine	
			[35], [37]
'paraffins'	1·53	'amines'	[135]
cyclohexane	0·43₃	'diazabicyclooctane'	
		in tetrachloromethane	[367]
trichloromethane	*1·5*	dimethyl methanamide	[209]
cyclohexane	1·182	dimethyl methanamide	[242]
tetrachloromethane	*0·948 ± 0·003ᵃ*	dimethyl methanamide	[157]
	0·930 ± 0·003ᵇ		
	0·910 ± 0·003ᶜ		
trichloromethane	*1·6*	dimethyl ethanamide	[209]
tetrachloromethane	*1·23ᵈ*	dimethyl ethanamide	[376]
	1·22ᵉ		
	1·19ᶠ		
	1·17ᵍ		
tetrachloromethane	*1·088 ± 0·003ᵃ*	diethyl ethanamide	[157]
	1·048 ± 0·003ᵇ		
	1·028 ± 0·003ᶜ		
tetrachloromethane	*1·050 ± 0·003ᵃ*	1-ethanoylperhydroazine	[157]
	1·027 ± 0·003ᵇ		
	1·000 ± 0·003ᶜ		
tetrachloromethane	*1·067 ± 0·003ᵃ*	1-methyl-1,2-dihydroazinone-2	
	1·015 ± 0·003ᵇ		[157]
	0·995 ± 0·003ᶜ		
tetrachloromethane	*0·801 ± 0·003ᵃ*	diethyl chloroethanamide	[157]
	0·767 ± 0·003ᵇ		
	0·735 ± 0·003ᶜ		
cyclohexane	*1·395 ± 0·010*	1-methylazolidinone-2	
		in cyclohexane	[400]
tetrachloromethane	*1·15ᵈ*	methyl ethanamide	[376]
	1·02ᵉ		
	0·92ᶠ		
	0·85ᵍ		
trichloromethane	*1·75*	azolidinone-2	[44]
'hydrocarbons'	0·81	propanone	[337]
'hydrocarbon'	0·53	propanone	[358]
cyclohexane	0·85	propanone	[182]
cyclohexane	0·897	propanone	[296]
cyclohexane	0·94 ± 0·01	propanone	[59]
cyclohexane	0·812	propanone	[242]
hexane	0·66	propanone	[136]
vapour	1·013	propanone	[392]
tetrachloromethane	0·90	propanone	[136]
trichloromethane	0·15	propanone	[207]
cyclohexane	*1·00 ± 0·01*	propanone	
		in cyclohexane	[59]
cyclohexane	*0·972 ± 0·006*	propanone	
		in cyclohexane	[400]
tetrachloromethane	*1·419 ± 0·014*	propanone	
		in tetrachloromethane	[180]
cyclohexane	*1·37*	propanone	[264]
hexane	0·78	butanone	[136]
hexane	0·80	pentanone-2	[136]
hexane	0·86	pentanone-3	[136]
hexane	0·96	cyclohexanone	[136]
cyclohexane	*0·984 ± 0·010*	cyclohexanone	
		in cyclohexane	[400]
hexane	0·71	3-methylbutanone-2	[136]

TABLE 3.11—*continued*

Initial medium	Chemical shift difference/ppm	Final medium	Ref.
hexane	0·80	4-methylpentanone-2	[136]
hexane	0·80	2,4-dimethylpentanone-3	[136]
cyclohexane	0·911	heptanone-4	[296]
cyclohexane	0·864	2,6-dimethylheptanone-4	[296]
hexane	0·86	2,6-dimethylheptanone-4	[136]
hexane	0·66	2,2,6,6-tetramethylheptanone-4	[136]
cyclohexane	0·851	nonanone-5	[296]
trichloromethane	*1·18*	propanone	[296]
'paraffins'	0·90	'ketones'	[135]
cyclohexane	*1·13*	1,3-dioxolanone-2	[264]
cyclohexane	*1·17*	4-methyl-1,3-dioxolanone-2	[264]
cyclohexane	0·40	ethanoic acid	[242]
'paraffins'	0·70	'ethers'	[135]
cyclohexane	*0·842 ± 0·009*	ethyl ethanoate *in cyclohexane*	[400]
'hydrocarbons'	0·76	diethyl ether	[337]
hexane	0·60	diethyl ether	[136]
cyclohexane	0·605	diethyl ether	[242]
cyclohexane	0·76	diethyl ether	[296]
cyclohexane	0·72	diethyl ether	[59]
tetrachloromethane	0·67	diethyl ether	[136]
trichloromethane	0·23	diethyl ether	[207]
cyclohexane	*0·905 ± 0·008*	diethyl ether *in cyclohexane*	[180]
cyclohexane	*0·82 ± 0·01*	diethyl ether *in cyclohexane*	[59]
tetrachloromethane	*1·266 ± 0·018*	diethyl ether *in tetrachloromethane*	[180]
cyclohexane	0·632	dipropyl ether	[296]
hexane	0·71	di(2-propyl) ether	[136]
cyclohexane	0·713	di(2-propyl) ether	[296]
tetrachloromethane	*1·126 ± 0·019*	di(2-propyl) ether	[180]
cyclohexane	0·502	dibutyl ether	[296]
cyclohexane	*0·89 ± 0·05*	dibutyl ether	[400]
hexane	0·41	di(2-methyl-2-propyl)ether	[136]
cyclohexane	0·587	1,4-dioxane	[296]
trichloromethane	0·28	1,4-dioxane	[207]
hexane	0·75	1,4-dioxane	[136]
heptane	*0·65*	1,4-dioxane *in heptane*	[193]
cyclohexane	*0·676 ± 0·006*	1,4-dioxane *in cyclohexane*	[400]
cyclohexane	0·68	oxane	[59]
cyclohexane	*0·83 ± 0·01*	oxane	[59]
hexane	0·90	oxolane	[136]
cyclohexane	0·900	oxolane	[296]
tetrachloromethane	0·73	oxolane	[136]
cyclohexane	*0·85 ± 0·03*	oxolane *in cyclohexane*	[96]
cyclohexane	0·76	oxetane	[59]

cyclohexane	0.87 ± 0.01	oxetane	
		in cyclohexane	[59]
cyclohexane	0.79	2-methyloxetane	[59]
cyclohexane	0.91 ± 0.01	2-methyloxetane	
		in cyclohexane	[59]
cyclohexane	0.64	oxirane	[59]
cyclohexane	0.76 ± 0.01	oxirane	
		in cyclohexane	[59]
cyclohexane	0.73	methyloxirane	[59]
cyclohexane	0.80 ± 0.01	methyloxirane	
		in cyclohexane	[59]
trichloromethane	0.95	1,3-dioxolanone-2	[264]
trichloromethane	0.98	4-methyl-1,3-dioxolanone-2	[264]
'paraffins'	0.87	'alcohols'	[135]
cyclohexane	0.611	methanol	[242]
hexane	0.76	methanol	[136]
hexane	0.88	ethanol	[136]
hexane	0.91	propanol-1	[136]
hexane	0.93	propanol-2	[136]
hexane	0.93	butanol	[136]
hexane	0.95	2-methylpropanol-2	[136]
hexane	1.13	cyclohexanol	[136]
hexane	0.71	4-methylpentanol-3	[136]
hexane	0.51	2,6-dimethylheptanol-4	[136]
cyclohexane	0.382	nitromethane	[242]
cyclohexane	0.499 ± 0.010	nitroethane	
		in cyclohexane	[400]
cyclohexane	0.410	nitrobenzene	[242]
cyclohexane	1.20	dimethyl sulphoxide	[263]
vapour	1.53_5	dimethyl sulphoxide	[392]
cyclohexane	1.50	dimethyl sulphoxide	[264]
trichloromethane	1.32	dimethyl sulphoxide	[264]
tetrachloromethane	0.943 ± 0.003^a	dimethyl sulphoxide	[157]
	0.945 ± 0.003^b		
	0.930 ± 0.003^c		
tetrachloromethane	1.078 ± 0.003^a	dibutyl sulphoxide	[157]
	1.043 ± 0.003^b		
	1.013 ± 0.003^c		
tetrachloromethane	1.158 ± 0.003^a	di(2-propyl) sulphoxide	[157]
	1.140 ± 0.003^b		
	1.127 ± 0.003^c		
tetrachloromethane	0.983 ± 0.003^a	thiolane 1-oxide	[157]
	0.972 ± 0.003^b		
	0.973 ± 0.003^c		
tetrachloromethane	0.607 ± 0.003^a	diethoxy sulphoxide	[157]
	0.598 ± 0.003^b		
	0.580 ± 0.003^c		
cyclohexane	0.45	trichlorophosphine oxide	[257]
cyclohexane	0.53	dichloro(methylthio)phosphine oxide	[257]
cyclohexane	0.63	dichloro(methoxy)phosphine oxide	[257]
cyclohexane	0.65	methyldifluorophosphine oxide	[257]
cyclohexane	0.67	methyldichlorophosphine oxide	[257]
cyclohexane	0.76	dichloro(dimethylamino)phosphine oxide	[257]

Field road, near Spring Ridge

sort he always welcomed, for spending time away from his verandah and store, family, and land.

No destination, however, beckoned to him with the same allure of liberty as the West. There, John Bass could be master of himself and Nature, free of the moral compromises and ambiguities that civilized life entails. Indeed, the images of my father's partners, ready with their rifles or repairing their broken-down car on some near-impassable desert road or holding up prey for the camera, disclose the essence of what John Bass found attractive about these trips to remote deserts and seashores and mountains. It was the *absence* of the binding ties of genealogy, freedom from the cultural pulls back to the land and a way of living on the land appropriate to his caste, imposed on him by the strong will of Sister Erin.

In Nature raw and untamed, the rigid hierarchical divisions between servile black and white master dissolves; my father especially liked taking pictures of this dissolution. They show the men, black and white, master and overseer, as a happy, savage band of brothers, pulling together to launch out in a boat, relaxing in camp by the fire, pretending to be "shipwrecked" on a beach, digging the Model-T Ford out of the sand.

On the hunting trips, as well, John Bass could do two other things he did much better than grow cotton: take pictures, and shoot. (The hand that pulls the trigger snaps the shutter, Roland Barthes reminds us; there is a connection between photography, which stills, and the rifle fire that fells and kills.) Like guns, cameras can kill, by excluding what must not, cannot be shown in the narrative constructed by all family photography. My father rarely took pictures of his family, and never, to my knowledge, allowed himself to be photographed with them. I know of no photograph of the entire Spring Ridge family—John Bass, his wife, Anne, his three children. The privilege of portrayal in his photography belonged to the males—his hunting buddies, his cousins and father (who performed his part as Southern gentleman with greater consistency than his son, but seems glad enough, in these pictures, to be away from it).

These photos—perhaps all photographs—are what they seem, and are *not* what they seem. If the explicit subject of these photographs is hunting, their background topic is the farm left far behind, recollected in the victorious smiles of men who had made good another escape from it. Similarly, John Bass's beautiful images of unpeopled landscapes—stands of saguaro cactus, the arid coast of the Gulf of California, the flats of Sonora—are finally not pictures of Nature. Anyone growing up in my family would immediately have known that each image was carefully composed according to the picturesque canons provided by the cardboard Stereopticon slides always available for viewing in my grandfather's house. John Bass's landscape photos recall the old imperialist vision of the vast and magnificent Western landscape: emptied of aboriginal peoples and frightening animals, "empty" space, yet full of significance, declaring to anyone who reads aright (as the social theorist Walter Benjamin teaches about such images) *that the photographer had been there*—that the occupation and pacification of the terrain by technology is now complete.

In the pacific, subdued emptiness, Bass found the peace that elsewhere eluded him all his life.

Of these photographs on my desk, my father is outside history, free from the linkages of genealogy, loosed from the Southern agrarian ideals of civility, Stoic morals, and devotion to the common good that I cherish, and he did not, perhaps because he could not.

Proudly, John Bass forever holds up the eagle he killed, Aleck slaughters and skins the buck, my grinning grandfather swings a dead bird by its neck—all beyond time, in the timeless space of photography. There, John Bass grasps the clutch of California quail (*we killed all we wanted*), rambles and hunts with his friends, drinks at Mr. Ren's club, sleeps it off, then rides again into the silence of the dry Mexican hills to hunt.

Had he lived, I might have become like him. He did not, and my soul's earliest formation became the business of others—Sister Erin and Aunt Vandalia, my mother's sister Aunt Antoinette, teachers and pastors—whose sharply differing views of what I should become (Methodist or Christian Scientist, small-town or urban, Southern or American) settled uneasily within a boy's mind already ill, and ill-at-ease. With the passage of years and inner troubles—long past the time when most men choose their course in life—I eventually found a way out of stagnation into a career, as husband, writer, dweller in great cities, that finally belonged to me.

It took a while longer for me to recognize the resemblance of my chosen path to that of my father—who, no better than I, could abide the Southern land, or stay within the old rhythms and strictures of settled Southern existence. Hence the peculiar role of his hunting album in my thinking, its uncertain place among the things that have come down to me from my family's past. If both my father and I found peace outside the South, I cannot renounce, as he did, the South's deepest understandings of the just, humane life, or the Classical and Southern ideals of civilized human harmony with the earth and under the sky, irreconcilable with the isolation and individualism symbolized for him by the desert.

Yet the album is a second mind to my own, a second heart—a book I cannot destroy, or give away to some archive of "family albums," or even cherish as a treasured memento from the past. It is almost all I shall ever know of my father, who left me two things. One was the land at Spring Ridge, from which I gathered the first inner topographies of my existence, my first understandings of what it is to be Southern in the twilight of rural Southern civilization. The other part of my inheritance is the collection of hunting photographs, which has traveled with me down the years and through all changes. In these pictures, I meet and know the man who engendered me toward the end of his brief, disappointed life, imprinting my mind and destiny on genealogy's unfurling banner of identities.

TABLE 3.11–*continued*

Initial medium	Chemical shift difference/ppm	Final medium	Ref.
cyclohexane	0·79	chlorodimethoxyphosphine oxide [257]	
tetrachloromethane	0·923[a] 0·865[b] 0·913[c]	trichloromethyldiethoxyphosphine oxide [156]	
cyclohexane	0·92	tris(methylthio)phosphine oxide [257]	
cyclohexane	0·940[a] 0·906[b] 0·878[c]	trichloromethyldiethoxyphosphine oxide [156]	
cyclohexane	1·00	dimethoxy(methylthio)phosphine oxide [257]	
cyclohexane	1·14	dimethoxyphosphine oxide	[257]
tetrachloromethane	1·155[a] 1·158[b] 1·160[c]	triethoxyphosphine oxide	[156]
tetrachloromethane	1·153[a] 1·140[b] 1·158[c]	methyldiethoxyphosphine oxide	[156]
cyclohexane	1·17	trimethoxyphosphine oxide	[257]
cyclohexane	1·23	triethoxyphosphine oxide	[257]
cyclohexane	1·330[a] 1·281[b] 1·270[c]	triethoxyphosphine oxide	[156]
cyclohexane	*1·339 ± 0·010*	triethoxyphosphine oxide in cyclohexane	[400]
cyclohexane	1·31	tripropoxyphosphine oxide	[257]
cyclohexane	1·28	tributoxyphosphine oxide	[257]
cyclohexane	*1·37*	tributoxyphosphine oxide	[289]
tetrachloromethane	1·393[a] 1·400[b] 1·423[c]	2-propyldiethoxyphosphine oxide [156]	
cyclohexane	1·31	methyldi(2-propyl)phosphine oxide [257]	
tetrachloromethane	1·415[a] 1·405[b] 1·383[c]	triphenylphosphine oxide	[156]
cyclohexane	1·41	dimethoxy(dimethylamino)-phosphine oxide	[257]
cyclohexane	1·43	methyldiethoxyphosphine oxide [257]	
cyclohexane	1·558[a] 1·525[b] 1·500[c]	2-propyldiethoxyphosphine oxide [156]	
tetrachloromethane	1·790[a] 1·802[b] 1·815[c]	tris(dimethylamino)phosphine oxide [156]	
cyclohexane	2·020[a] 1·998[b] 1·998[c]	tris(dimethylamino)phosphine oxide [156]	
cyclohexane	2·03	tris(dimethylamino)phosphine oxide [257]	
cyclohexane	1·995	tris(dimethylamino)phosphine oxide [242]	

cyclohexane	*2·02*	tris(dimethylamino)phosphine oxide	[291]
dichloromethane	*7·41*(?)	tris(dimethylamino)phosphine oxide	[291]
cyclohexane	*2·03*	trioctylphosphine oxide	[289]
cyclohexane	0·35	methyldichlorophosphine sulphide	[257]
cyclohexane	0·47	tris(methylthio)phosphine sulphide	[257]
cyclohexane	0·59	trimethoxyphosphine sulphide	[257]
cyclohexane	0·62	methyldi(methoxy)phosphine sulphide	[257]
cyclohexane	0·69	methyldi(2-propyl)phosphine sulphide	[257]
cyclohexane	0·73	methyl di(2-propyl)phosphine sulphide	[257]
cyclohexane	1·10	tris(dimethylamino)phosphine sulphide	[257]
cyclohexane	*0·579 ± 0·006*	dimethyl sulphide *in cyclohexane*	[400]
cyclohexane	0·61	diethyl sulphide	[59]
cyclohexane	*0·933 ± 0·005*	diethyl sulphide *in cyclohexane*	[400]
cyclohexane	*0·75 ± 0·01*	diethyl sulphide *in cyclohexane*	[59]
cyclohexane	0·33	thiirane	[59]
cyclohexane	*0·38 ± 0·01*	thiirane *in cyclohexane*	[59]
cyclohexane	0·76	thietane	[59]
cyclohexane	*0·79 ± 0·01*	thietane *in cyclohexane*	[59]
cyclohexane	0·68	thiolane	[59]
cyclohexane	*0·83 ± 0·01*	thiolane *in cyclohexane*	[59]
cyclohexane	*0·873 ± 0·006*	thiolane *in cyclohexane*	[400]
cyclohexane	0·65	thiane	[59]
cyclohexane	*0·74 ± 0·01*	thiane *in cyclohexane*	[59]
cyclohexane	0·70	trimethoxyphosphine selenide	[257]
cyclohcxane	0·38	tris(methylthio)phosphine	[257]
cyclohexane	0·60	trimethoxyphosphine	[257]
cyclohexane	0·66	tris(dimethylamino)phosphine	[257]
trichloromethane	*0·75*	cyclohexyl isocyanide	[36]
trichloromethane	*0·76*$_2$	cyclohexyl isocyanide	[37]
tetrachloromethane	*2·92 ± 0·12*	tetrabutylammonium chloride *in tetrachloromethane*	[158]
ethanenitrile	*3·08 ± 0·11*	tetrabutylammonium chloride *in ethanenitrile*	[158]
tetrachloromethane	*2·51 ± 0·08*	tetrabutylammonium bromide *in tetrachloromethane*	[158]
ethanenitrile	*2·76 ± 0·33*	tetrabutylammonium bromide *in ethanenitrile*	[158]
tetrachloromethane	*2·30 ± 0·90*	tetraheptylammonium iodide *in tetrachloromethane*	[158]
ethanenitrile	*2·18 ± 0·22*	tetrabutylammonium iodide *in ethanenitrile*	[158]

D·

TABLE 3.12

Predicted and observed solvent shifts of trichloromethane from cyclohexane to donor solvents

Solvent	Δ_s/ppm calculated	Δ_s/ppm observed
trichloromethane	0·21	0·15–0·29
benzene	−0·98	−0·75– −1·35
propanone	0·67	0·81–0·90
diethyl ether	0·71	0·60–0·76
oxolane	0·72	0·900
triethylamine	1·22	0·967–1·20

TABLE 3.13

Carbon-13 NMR parameters of $HCCl_3$ complexes

Solvent	$\Delta_s{}^a$/ppm	$^1J_{CH}{}^b$/Hz
cyclohexane	0·00	208·1; 208·11[c]
tetrachloromethane	0·20	208·4; 208·26[c]
trichloromethane	0·37	209·5; 208·91[c]
benzene	0·47	210·6
ethanoic acid	1·02	
ethanoyl chloride		211·8
nitromethane	1·12	213·6
nitrobenzene	1·12	
diethyl ether	1·24	213·7
methanol	1·35	214·3
ethanenitrile	1·63	214·6
propanone	1·76	215·2
triethylamine	1·85	214·2
dimethyl methanamide	2·55	217·4; 216·46[c]
azine	2·63	215·0[d]
dimethyl sulphoxide		217·7
tris(dimethylamino)phosphine oxide	4·24	

[a] Ref. 242; [b] Ref. 116; [c] Ref. 391; [d] Ref. 235.

TABLE 3.14

Dipole moments of complexes involving trichloromethane (A) and a Donor (B)

Solvent	Donor	Complex	Dipole moment 10^{30} μ/C m	Method	Ref.
—	diethyl ether	AB	6·87	a	[113]
—	diethyl ether	AB	8·67	a	[18]
c-C_6H_{12}	diethyl ether	AB	7·20[d]; 7·10[e]	b	[201]
CCl_4	diethyl ether	AB	8·64[d]; 8·87[e]	b	[201]
—	di(2-propyl) ether	AB	7·57	a	[113]
CCl_4	di(2-propyl) ether	AB	8·47[d]; 8·74[e]	b	[201]

c-C_6H_{12}	oxolane	**AB**	8·44	a	[393]
c-C_6H_{12}	oxolane	**A_2B**	10·17	a	[393]
—	1,4-dioxane	**AB**	4·80	a	[18]
—	1,4-dioxane	**A_2B**	4·27	a	[18]
—	propanone	**AB**	12·0	a	[18]
CCl_4	propanone	**AB**	13·0	a	[100]
CCl_4	propanone	**AB**	13·24[d]; 13·01[f]	b	[201]
—	propanone	**A_2B**	14·0	a	[18]
CCl_4	azine	**AB**	12·94	b	[201]
CCl_4	ethanenitrile	**AB**	17·08	b	[201]
—	triethylamine	**AB**	10·5$_1$	b	[214]
c-C_6H_{12}	triethylamine	**AB**	6·90	b	[201]
—	benzene	**AB**	4·24	c	[46]
—	tetrachloromethane	**(AB)**	4·00	c	[46]
—	heptane	**(AB)**	4·00	c	[46]

[a] Dielectric polarisation measurements.
[b] Dielectric constant measurements.
[c] Infrared spectroscopic measurements.
[d] Assuming C–H bond at 0° to symmetry axis of the donor molecule.
[e] Assuming C–H bond at 55° to symmetry axis of the donor molecule.
[f] Assuming C–H bond at 60° to symmetry axis of the donor molecule.

TABLE 3.15

Stoichiometry of higher complexes involving trihalomethane(**A**) and an electron donor(**B**)

Donor	Complex	Method of Detection	Ref.
	1. $HCBr_3$		
dimethyl ethanamide	**A_2B**	infrared spectroscopy	[83]
triethylphosphine	**AB_3**	elemental analysis	[327]
tetramethyl urea	**A_2B**	infrared spectroscopy	[83]
1-methylazolidinone-2	**A_2B**	infrared spectroscopy	[83]
'propionopiperidide'	**A_2B**	infrared spectroscopy	[83]
	2. HCF_3		
tetraethylammonium chloride	**A_2B**		[103]
	A_4B	pressure/composition	
	3. $HCFCl_2$		
2,5,8,11,14-pentaoxapentadecane	**A_3B**	enthalpy of mixing	[406]
	4. HCl_3		
triethylphosphine	**AB_3**	elemental analysis	[327]
1-azanaphthalene	**AB_3**	elemental analysis of crystal	[327]
1-azanaphthalene	**AB_3**	X-ray crystal structure	[42], [43]
sulphur, S_8	**AB_3**	X-ray crystal structure	[41], [43]
1,4-diselenane	**A_2B**	X-ray crystal structure	[170]

FREEING THE WATERS

Cemetery, Henderson

IF THE DEATH OF AUNT VANDALIA HAD MARKED THE ONSET OF A LONG season of remembering, it also began a time of letting go. As I slowly paced up the walkway to the front door of my aunt's house the day after she died, my thoughts were turning toward this impending snap of the last strong historical link I had with the South and the Southern ground I'd left years before. My sister and I had known for years that, once the old lady was in the ground, and once all we treasured among her possessions had been scooped up and carted off, the house and its ground would have to be sold. In the late 1960s, I had imagined going back to live, to write in the lovely isolation of Greenwood, all my days. By 1990, this dream from a troubled time in my life had gradually died a good and peaceful death.

To sell the house would be to doom our homeplace to almost certain demolition. But if that was its fate, then so it would have to be. Neither my sister nor I wanted to live there, for even a fraction of the year. One of my own reasons was that, whatever isolation Greenwood had once enjoyed from the industrialized South—the quiet Mayesville still enjoyed, along with many other little Southern towns—it was now utterly abolished by the sprawl of suburbia along Interstate 20 from nearby Shreveport, and the incessant roar of the highway itself. A muddle of cheap, ugly townhouses now stood just down from the house in once-empty pastures where I'd chased rabbits as a boy, and a noisy, dog-ridden trailer court had crowded up to within a whisper's distance of the screened back porch where my grandparents, aunt and uncle, and I had taken breakfast on summer mornings.

Then there were the practical circumstances, which caused my only Mays kinsman then in need of a dwelling, cousin Lane, to decide he had no use for the old place. To halt the encroachment of weed and ruin on the grounds—to pull the gardens and lawns back to the merest presentability from tangled waste—would involve a large investment of time and money. Indeed, unless a great deal of cash and energy were put in, the old gardens and yards would never

TABLE 3.16

Formation constants of complexes containing trihalomethane

Donor	T/K	K	Units	Solvent	Ref.
1. HCF$_3$					
oxolane	264·9	7·6$_0$	mf^{-1}	c-C$_6$H$_{12}$	[96]
	266·7	7·8$_6$	mf^{-1}	c-C$_6$H$_{12}$	[96]
	267·5	8·00	mf^{-1}	c-C$_6$H$_{12}$	[95]
	274·4	6·86	mf^{-1}	c-C$_6$H$_{12}$	[95]
	276·1	6·1$_7$	mf^{-1}	c-C$_6$H$_{12}$	[96]
	280·4	5·8$_3$	mf^{-1}	c-C$_6$H$_{12}$	[96]
	280·7	6·24	mf^{-1}	c-C$_6$H$_{12}$	[95]
	282·1	5·3$_3$	mf^{-1}	c-C$_6$H$_{12}$	[96]
	290·3	5·35	mf^{-1}	c-C$_6$H$_{12}$	[95]
	298·8	4·70	mf^{-1}	c-C$_6$H$_{12}$	[95]
	299·0	4·80	mf^{-1}	c-C$_6$H$_{12}$	[95]
	299·6	4·70	mf^{-1}	c-C$_6$H$_{12}$	[95]
	299·9	4·9$_3$	mf^{-1}	c-C$_6$H$_{12}$	[96]
	300·9	4·5$_6$	mf^{-1}	c-C$_6$H$_{12}$	[96]
	309·4	4·09	mf^{-1}	c-C$_6$H$_{12}$	[95]
	311·2	3·91	mf^{-1}	c-C$_6$H$_{12}$	[95]
	311·5	3·5$_7$	mf^{-1}	c-C$_6$H$_{12}$	[96]
2. DCF$_3$					
oxolane	267·5	8·41	mf^{-1}	c-C$_6$H$_{12}$	[95]
	274·4	7·34	mf^{-1}	c-C$_6$H$_{12}$	[95]
	290·3	5·55	mf^{-1}	c-C$_6$H$_{12}$	[95]
	298·8	5·00	mf^{-1}	c-C$_6$H$_{12}$	[95]
	299·0	4·85	mf^{-1}	c-C$_6$H$_{12}$	[95]
	299·6	4·87	mf^{-1}	c-C$_6$H$_{12}$	[95]
	309·4	4·21	mf^{-1}	c-C$_6$H$_{12}$	[95]
	311·2	4·05	mf^{-1}	c-C$_6$H$_{12}$	[95]
	319·0	3·63	mf^{-1}	c-C$_6$H$_{12}$	[95]
	322·7	3·50	mf^{-1}	c-C$_6$H$_{12}$	[95]
3. HCF$_2$Cl					
didecyl benzene-1,2-dicarboxylate	298	4·98	mf^{-1}	–	[347]
dicapryl benzene-1,2-dicarboxylate	298	4·49	mf^{-1}	–	[347]
dioctyl benzene-1,2-dicarboxylate	298	4·43	mf^{-1}	–	[347]
dioctyl decanedioate	298	3·87	mf^{-1}	–	[347]
1,2,3-tris(methylcarbonyloxy)-propane	298	7·69	mf^{-1}	–	[347]
dimethyl methanamide	298	17·1	mf^{-1}	–	[347]
4. HCCl$_2$Br					
propanone	304	0·80 ± 0·05	M^{-1}	C$_6$H$_{14}$	[203]
dioctyl ether	303·15	0·407 ± 0·010	M^{-1}	–	[352]
	313·15	0·363 ± 0·010	M^{-1}	–	[352]
	323·15	0·324 ± 0·010	M^{-1}	–	[352]
	333·15	0·288 ± 0·010	M^{-1}	–	[352]
dioctyl sulphide	303·15	0·501 ± 0·010	M^{-1}	–	[352]
	313·15	0·454 ± 0·010	M^{-1}	–	[352]
	323·15	0·411 ± 0·010	M^{-1}	–	[352]
	333·15	0·374 ± 0·010	M^{-1}	–	[352]
5. HCCl$_2$CN					
benzene	313	0·31	M^{-1}	–	[193]
methylbenzene	313	0·38	M^{-1}	–	[193]

present to the world the tidy order and rhythmic beauty militantly enforced on that patch of Louisiana hilltop decades before by Sister Erin, Aunt Vandalia, and their gardening help. And what would be the point of going to all that trouble to restore a place to its former delight when the site itself was hemmed in by sprawl, dogs, traffic noise?

As for the house itself—it would surely be unacceptable to most any Southerner I knew who had come of age since the Second World War, to Lane, certainly to me.

My grandmother had designed the structure her husband then built with the torrid summer heat in mind, and made ample provision for shade by creating broad overhanging eaves and wide porches. And she kept her house architecturally porous, by inserting windows broad and tall, and double doors opening onto porticos and between rooms. This was all, of course, long before the 1950s, when windows across the South were slammed shut, the porches emptied, the usual after-dinner visiting stopped, because of air-conditioning. It was as though everyone had suddenly waked from a long sleep, realized how unpleasantly warm our summer evenings were—how tiresome those traditional evening visits to kin and friends had become—then retreated inside to huddle in refrigerated air before their television sets.

The very openness Sister Erin intended for keeping the interior cool, breezy, and inviting throughout the long subtropical summer made conversion to air-conditioning impractical. And in winter, any occupant would have to put up with the stuffiness and trouble of gas fires and log-burning hearth, chilly drafts whistling up through floorboards and squeezing through cracks, pipes kept exposed under the house to keep the water cool in summer, and, for that reason, liable to freeze and burst on the coldest nights. All these facts added up to a legion of quite sound reasons for not keeping the house. Like the people in it, the gracious old frame structure had grown old, outlived the small-town Southern culture that shaped its design; and now its time was done.

The sale of the house, I imagined while walking slowly up to the front door that spring morning, would be the last snip to binding entwinements; and, until that final cut was made, I would have nothing but the routine work of burial and legal matters to attend to.

But more immediate farewells waited for me on the other side of the dark front door—leave-takings of a sort that had slipped my mind, because I had lived so long beyond the reach of the old codes that, even as late as 1990, still governed the rites of Southern death and burial.

All within had been readied for my arrival, the bearing-off of Aunt Vandalia's body seen to by household staff, and her bedclothes replaced by tautly ironed linens scented with lavender. The women, all of them white, who had served her decided our initial greeting was to be in the still-dark sickroom—not, as I would have imagined, given the warm weather, in the living room or on the pleasant south verandah. They had left virtually everything as it had been at the time of Vandalia's death, somber and unkempt. The paper shades were down and secured by hooks, the curtains on the windows over-looking the lost rose garden were drawn, as they had been since the evening light failed the night she died. The huge oak rocker still stood in front of a tall, carved cabinet piled high with storybooks and toys from my aunt's childhood, and my father's. The robust reek of rub-bing alcohol and stronger antiseptics still hung in the closed air of the room, mixed with odors of medicines, and a sour stench of urine, and the sweet sift of old face powders. These kindly white servants—replacements of Aunt Vandalia's black ones, who had grown too old to work, or had moved away—waited in half-light tinted by decay, curtained off from the world's youth and health; so there we met.

Their condolences and my thanks were exchanged, and tales told me of Miss Vannie's saintly courage in suffering and Christian sweet-ness to the end. I listened to these formulaic tributes with the respect due them. For though outside the South for many years, I had not forgotten what to do: listen attentively, and with gratitude, to the hagiological lies servants tell of the dead. In a sense, the nurse had told me too much, and breached decorum, by truthfully relating my aunt's last sentiments. But there would have been no point in asking more about Aunt Vandalia's last days, what she said, or who she greeted the impending night. In dying, Vandalia had invited her own shrouding in blessed memory, her servants' last gift to her. And *of course*, I agreed with her erstwhile nurses: Aunt Vandalia's was an exemplary Christian passage from this vale of tears to the inner courts of glory. What other sort of passage could it be for a Southern

1,3,5-trimethylbenzene	313	0·70	M^{-1}	–	[193]
pentamethylbenzene	313	1·5	M^{-1}	–	[193]
1-methylnaphthalene	313	0·53	M^{-1}	–	[193]
1,4-dioxane	313	1·4	M^{-1}	–	[193]
		6. $HCClBr_2$			
dioctyl ether	303·15	0·416 ± 0·010	M^{-1}	–	[352]
	313·15	0·363 ± 0·010	M^{-1}	–	[352]
	323·15	0·318 ± 0·010	M^{-1}	–	[352]
	333·15	0·275 ± 0·010	M^{-1}	–	[352]
dioctyl sulphide	303·15	0·615 ± 0·010	M^{-1}	–	[352]
	313·15	0·555 ± 0·010	M^{-1}	–	[352]
	323·15	0·503 ± 0·010	M^{-1}	–	[352]
	333·15	0·457 ± 0·010	M^{-1}	–	[352]
		7. $HCBr_3$			
dioctyl ether	303·15	0·411 ± 0·010	M^{-1}	–	[352]
	313·15	0·370 ± 0·010	M^{-1}	–	[352]
	323·15	0·332 ± 0·010	M^{-1}	–	[352]
	333·15	0·297 ± 0·010	M^{-1}	–	[352]
oxolane	276·1	6·66	mf^{-1}	$c\text{-}C_6H_{12}$	[96]
	288·5	5·59	mf^{-1}	$c\text{-}C_6H_{12}$	[96]
	306·2	4·09	mf^{-1}	$c\text{-}C_6H_{12}$	[96]
	310·2	4·19	mf^{-1}	$c\text{-}C_6H_{12}$	[96]
	317·5	3·62	mf^{-1}	$c\text{-}C_6H_{12}$	[96]
	328·8	3·14	mf^{-1}	$c\text{-}C_6H_{12}$	[96]
diethyl ether	293	0·43–0·65	mf^{-1}	–	[113]
	293	0·44	mf^{-1}	–	[145]
di(2-propyl) ether	293	0·49–0·81	mf^{-1}	–	[113]
	293	0·49	mf^{-1}	–	[145]
propanone	304	0·45 ± 0·05	M^{-1}	C_6H_{14}	[203]
dioctyl sulphide	303·15	0·733 ± 0·010	M^{-1}	–	[352]
	313·15	0·653 ± 0·010	M^{-1}	–	[352]
	323·15	0·583 ± 0·010	M^{-1}	–	[352]
	333·15	0·521 ± 0·010	M^{-1}	–	[352]
tribromomethane	–	0·54 ± 0·10	mf^{-1}	CCl_4	[278]
	–	0·72 ± 0·24	mf^{-1}	$c\text{-}C_6H_{12}$	[278]
	298	0·62	M^{-1}	CCl_4	[359]
	313	0·57	M^{-1}	CCl_4	[359]
azine	300	0·34 ± 0·05	M^{-1}	CCl_4	[158]
tetrabutylammonium chloride	300	2·45 ± 0·16	M^{-1}	CCl_4	[158]
	248	1·64 ± 0·11	M^{-1}	CH_3CN	[158]
	273	1·32 ± 0·08	M^{-1}	CH_3CN	[158]
	298	1·09 ± 0·06	M^{-1}	CH_3CN	[158]
	323	0·99 ± 0·05	M^{-1}	CH_3CN	[158]
	348	0·85 ± 0·09	M^{-1}	CH_3CN	[158]
tetrabutylammonium bromide	300	1·86 ± 0·09	M^{-1}	CCl_4	[158]
	248	1·63 ± 0·16	M^{-1}	CH_3CN	[158]
	273	1·27 ± 0·15	M^{-1}	CH_3CN	[158]
	298	1·10 ± 0·14	M^{-1}	CH_3CN	[158]
	323	0·95 ± 0·12	M^{-1}	CH_3CN	[158]
	348	0·87 ± 0·12	M^{-1}	CH_3CN	[158]
tetraheptylammonium iodide	300	1·95 ± 0·45	M^{-1}	CCl_4	[158]
tetrabutylammonium iodide	248	1·58 ± 0·11	M^{-1}	CH_3CN	[158]
	273	1·29 ± 0·10	M^{-1}	CH_3CN	[158]
	298	1·06 ± 0·09	M^{-1}	CH_3CN	[158]
	323	1·13 ± 0·09	M^{-1}	CH_3CN	[158]
	348	1·01 ± 0·09	M^{-1}	CH_3CN	[158]
		8. HCI_3			
tetrabutylammonium chloride	300	3·30 ± 0·25	M^{-1}	CH_3CN	[158]

TABLE 3.16—*continued*

Donor	T/K	K	Units	Solvent	Ref.
tetrabutylammonium bromide	300	3.76 ± 0.22	M^{-1}	CH_3CN	[158]
tetrabutylammonium iodide	300	3.56 ± 0.18	M^{-1}	CH_3CN	[158]
azine	300	0.39 ± 0.07	M^{-1}	CCl_4	[158]
oxolane	276.1	6.30	mf^{-1}	$c\text{-}C_6H_{12}$	[96]
	299.9	5.18	mf^{-1}	$c\text{-}C_6H_{12}$	[96]
	300.2	5.02	mf^{-1}	$c\text{-}C_6H_{12}$	[96]
	311.1	4.46	mf^{-1}	$c\text{-}C_6H_{12}$	[96]
	320.0	4.35	mf^{-1}	$c\text{-}C_6H_{12}$	[96]
	326.8	3.80	mf^{-1}	$c\text{-}C_6H_{12}$	[96]
9. $HC(NO_2)_3$					
benzene	306.6	8.46	mf^{-1}	$c\text{-}C_6H_{12}$	[179]
methylbenzene	306.6	14.63	mf^{-1}	$c\text{-}C_6H_{12}$	[179]
1,2-dimethylbenzene	306.6	26.48	mf^{-1}	$c\text{-}C_6H_{12}$	[179]
1,3-dimethylbenzene	306.6	26.09	mf^{-1}	$c\text{-}C_6H_{12}$	[179]
1,4-dimethylbenzene	306.6	24.12	mf^{-1}	$c\text{-}C_6H_{12}$	[179]
1,3,5-trimethylbenzene	306.6	37.88	mf^{-1}	$c\text{-}C_6H_{12}$	[179]
1,2,4,5-tetramethylbenzene	306.6	78.11	mf^{-1}	$c\text{-}C_6H_{12}$	[179]
pentamethylbenzene	306.6	225.25	mf^{-1}	$c\text{-}C_6H_{12}$	[179]
hexamethylbenzene	306.6	279.40	mf^{-1}	$c\text{-}C_6H_{12}$	[179]

TABLE 3.17

Enthalpies of mixing of trihalomethanes with donor molecules

Donor	T/K	$\Delta_m H/J\ mol^{-1}$	Ref.
$HCFCl_2$			
propanone	273	-1840	[231]
diethyl ether	273	-2510	[231]
1,2-dimethoxyethane	276	-4810	[406]
2,5,8,11,14-pentaoxapentadecane	276	$-5650\ (\mathbf{A_3B})$	[406]
$HCBr_3$			
benzene	298	$-\ 142$	[377]
methylbenzene	288–93	$-\ 330$	[21]
	298	$-\ 393$	[377]
1,3-dimethylbenzene	298	$-\ 444$	[377]
1,3,5-trimethylbenzene	298	$-\ 452$	[377]
azine	283	-2085	[259]
dimethyl ethanamide	283	-3840	[259]
diethyl ether	283	-2155	[259]
propanone	283	-1430	[259]

TABLE 3.18

Enthalpies of formation of trihalomethane complexes

Donor	T/K	$\Delta_\phi H/kJ\ mol^{-1}$	$\dfrac{\Delta_\phi S}{J\ K^{-1}\ mol^{-1}}$	Solvent	Ref.
	1. HCF_3				
oxolane	300	$-10.0\ \pm 0.8$		$c\text{-}C_6H_{12}$	[96
	298	-10.85 ± 0.10		$c\text{-}C_6H_{12}$	[95
	[$+DCF_3$] 298	-11.41 ± 0.09		$c\text{-}C_6H_{12}$	[95

raethylammonium loride (A$_2$B)	195	-31.8		solid	[103]
2. HCF$_2$Cl					
ecyl nzene-1,2-dicarboxylate	298	-13.9		–	[347]
apryl nzene-1,2-dicarboxylate	298	-12.9		–	[347]
octyl nzene-1,2-dicarboxylate	298	-12.0		–	[347]
octyl decanedioate	298	-9.9		–	[347]
,3-tris(methylcarbonyloxy)propane	298	-11.7		–	[347]
methyl methanamide	298	-14.7		–	[347]
3. HCCl$_2$Br					
opanone	304	-13.8 ± 0.8		C$_6$H$_{14}$	[203]
ctyl ether	303.15–333.15	-11.5_5	-45.2	–	[352]
ctyl sulphide	303.15–333.15	-8.3_3	-31.4	–	[352]
4. HCCl$_2$CN					
nzene	313	-7.1	-32.6	–	[193]
,5-trimethylbenzene	313	-9.6	-34.3	–	[193]
methylnaphthalene	313	-6.3	-25.1	–	[193]
-dioxane	313	-14.6	-49.4	–	[193]
5. HCClBr$_2$					
ctyl ether	303.15–333.15	-9.6_7	-39.3	–	[352]
ctyl sulphide	303.15–333.15	-8.20	-32.8	–	[352]
6. HCBr$_3$					
ctyl ether	303.15–333.15	-9.08	-37.3	–	[352]
ctyl sulphide	303.15–333.15	-9.54	-34.1	–	[352]
romomethane	298–313	-4.2		CCl$_4$	[359]
olane	300	-10.9 ± 0.8		c-C$_6$H$_{12}$	[96]
opanone	304	-11.3 ± 0.8		C$_6$H$_{14}$	[203]
abutylammonium chloride	300	-4.60 ± 0.17		CH$_3$CN	[158]
abutylammonium bromide	300	-4.48 ± 0.21		CH$_3$CN	[158]
abutylammonium ide	300	-2.97 ± 0.38		CH$_3$CN	[158]
7. HCI$_3$					
olane	300	-6.7 ± 1.7		c-C$_6$H$_{12}$	[96]

TABLE 3.19

Solubilities of HCCl$_2$F at 305.4 K [407]
The 'Ideal' solubility is 0.381 mol mol^{-1}

Solvent	Solubility mol mol^{-1}
phenyl methyl ether	0.498
2,5-dioxahexane [1,2-dimethoxyethane]	0.576
1,5-dichloro-3-oxapentane	0.451
2,5,8-trioxanonane	0.611
3,6,9-trioxaundecane	0.640
1-chloro-3,6-dioxaheptane	0.486
2,5,8,11-tetraoxadodecane	0.638

woman of her sort, to what other destination than a seat at the hand of the Father?

It was after these ritual honors to my aunt that the encounter in half-light could have become awkward. I might have wondered: Why the sidewise glances, the nervous shifting from foot to foot, the twisting of damp handkerchiefs? But, curiously—or so it seemed to me at the time—I remembered how to read the eyes of servants. They appealed for directions about what to do next, for household instructions, for the opportunity to perform some service for me—if only to bring a glass of lemonade—for assurance that unpaid wages would indeed be paid. The looks told me I was now the master of the house, and that I had accepted this role.

There was nothing mystical about this effortless, nearly thought-less answering of tradition's call. From infancy on my father's planta-tion, I had been surrounded by house servants and had known field hands, been nursed and looked after by a black mammy, Essie De, undergone purposeful, instructive mixings with all sorts of whites, and allowed a limited mingling with blacks. From the earliest time I can remember, the white elders, people of various grandeurs and social positions—a wide range of men and women requiring exquisitely tuned and differing deference—passed through the parlor of my father, and along the south verandah of my grandfather. And the ease with which one took his or her place on the verandah, I learned, was a good measure of the person's proximity to power in this Louisiana village.

The only people continually circulating around us who had no proximity at all were the blacks. As Essie De has reminded me, Sister Erin and my grandfather John kept the blacks that surrounded them very firmly in their place—and, here, I am using the word in its most restricted, restricting sense. No black man or woman was allowed to come closer to the house than the back steps, other than the cook Peggy and other house servants, and Essie De, when she traveled up from Spring Ridge to look after me. They were not allowed to use a toilet in the house, and went instead to an outhouse Sister Erin had constructed for them at the bottom of the back garden. As far as I could tell, however, my grandparents were considered good employ-ers by Peggy, their splendid cook of some forty years, by the chauf-feur who drove Sister Erin and Aunt Vandalia to shop in Shreveport and exchange polite gossip at meetings of Greenwood's social clubs, by gardeners and field hands and other servants whose names I do not remember.

And they were loyal, and expressed that loyalty in ways that are somewhat hard to understand, given their strict exclusion from the inner culture of the family they served. When staying before Uncle Alvin's burial, in 1989, I heard a faint rapping on the door of the screened back porch, and answered it. The elderly black visitor—hat in hand, his eyes downcast, long after such deference of blacks toward whites had become, I imagined, obsolete—told me that he had once worked for Mr. Alvin, and wondered if there was anything he might do. I had never met this man, or heard his name; but as his story unfolded, it dawned on me that his employment by my uncle had begun and ended sixty years before. Despite the years, however, upon hearing of Alvin's death he had remembered some kindness done him long before, and come to repay it.

In my younger years, I learned, one might accept with gratitude such a gift from a black stranger, but could never expect to know the one who offered it. The codex of class and color imposed on me at birth forbade intimacy with blacks—apart from, of course, tenderness of the sort that prevailed between white children and their mammies, and the sexual transactions between white masters and black servants that persisted, well known but never mentioned among the district's white families, down to my own day in the plantation culture of the South. (Essie De, who has always had an impish sense of humor about country matters, likes to recall the unbearable snobbery of the comfortably kept black mistresses in the vicinity of Spring Ridge, and their haughty disdain for other black women, whom they considered *ordinary*.) I was permitted by my mother to play and go fishing with Essie De's young son, who was almost exactly my contemporary; but forbidden, when I asked, to stay the night at his house.

I grew up surrounded by blacks, as my ancestors did, without knowing them. There were many whites who worked in and around my family's dwelling places of whom I was similarly aware, but never knew. Because of some provision in the unwritten code of my caste, the place of these white laborers, housewives, and even small merchants—like that of all blacks—lay outside the circle of people I

TABLE 3.19—*continued*

Solvent	Solubility mol mol^{-1}
1-chloro-3,6,9-trioxadecane	0·578
2,5,8,11,14,17-hexaoxaheptadecane	0·706
2,5,8,11,14,17,20-heptaoxaheneicosane	0·716
1,7-dichloro-4-oxaheptane	0·421
2,4-di(chloromethyl)-3-oxapentane	0·425
4-fluorophenyl methyl ether	0·385
1,3-diethoxybenzene	0·456
1,4-dioxane	0·479
2-(2,5-dioxanonyl)oxolane	0·608
1,9-dioxalanyl-2,5,8-trioxanonane	0·680
diphenyl sulphide	0·306
2,5,8-trithianonane	0·448
butyl butanoate	0·546
ethyl dodecanoate	0·556
diethyl benzene-1,2-dicarboxylate	0·543
diethyl ethanedioate	0·545
diethyl propane-1,3-dioate	0·558
1,2,3-tris(methylcarbonyloxy)propane	0·570
1,2,3-tris(ethylcarbonyloxy)propane	0·609
1,2,3-tris(propylcarbonyloxy)propane	0·573
1,2,3-tris(pentylcarbonyloxy)propane	0·661
1,3-bis(methylcarbonyloxy)propane	0·568
1,3-dichloro-2-propyl ethanoate	0·397
1,3-dimethylbutyl ethanoate	0·542
3,6-dioxaoctanone-2	0·542
3,6,11,14-tetraoxahexadecane-7,10-dione	0·624
3,6,13,16-tetraoxaoctadecane-7,12-dione	0·662
4-phenyl-3-oxabutyl ethanoate	0·538
bis(3-oxabutyl) benzene-1,2-dicarboxylate	0·582
bis(3-oxaheptyl) benzene-1,2-dicarboxylate	0·631
3,6,9,12-tetraoxatetradecane-5,10-dione	0·602
3,6,9,12,15-pentaoxapentadecane-5,13-dione	0·634
2,5,8,11,14-pentaoxatridecane-4,9-dione	0·59
4-oxolanyl-3-oxabutyl ethanoate	0·598
2,5,8,11,14,17-hexaoxaoctadecane-4,15-dione	0·636
2,5,8,11,14-pentaoxahexadecane-4,15-dione	0·681
3,6,9,12-tetraoxatetradecane-2,13-dione	0·631
3,6,9,12-tetraoxatridecan-10-one	0·620
3,6-dioxadecyl ethanoate	0·587
3,6,9-trioxaundecane-2,10-dione	0·599
3-oxaheptyl butanoate	0·565
3-oxaheptyl ethanoate	0·562
3-oxaheptyl dodecanoate	0·594
3,6-dioxaheptyl ethanoate	0·596
3,6,9-trioxadecyl ethanoate	0·632
oxolanylmethyl dodecanoate	0·610
oxolylmethyl ethanoate	0·438
ethyl oxole-2-carboxylate	0·482
2,5,9,12-tetraoxatridecane-4,10-dione	0·587
oxolanylmethyl 3-oxabutanoate	0·586
oxolanylmethyl benzoate	0·524
2,5,7,10-tetraoxaundecan-6-one	0·540
3-oxanonane-4,7-dione	0·566

5- methyl-2-(2-propyl)cyclohexanone	0·545
2-formyloxole	0·367
3-oxapentyl 2-hydroxyethanoate	0·423
(hydroxymethyl)oxolane	0·368
4-phenyl-3-oxabutan-1-ol	0·368
3,6-dioxaoctane-1,8-diol	0·330
3-oxapentane-1,5-diol	0·288
1,3-propanediol	0·073
ethane-1,2-diol	0·055
aminobenzene	0·258
(dimethylamino)benzene	0·425
4-aminophenyldimethylamine	0·384
1-azanaphthalene	0·443
3,6-diethyl-3,6-diazaoctane-4,5-dione	0·651
ethyl 4-ethyl-4-aza-3-hexanoate	0·633
ethanoyl perhydroazine	0·618
nitrobenzene	0·360
1-fluoronaphthalene	0·317
(trifluoromethyl)benzene	0·323
bicyclo[4.4.0]deca-2,4-diene	0·379
bicyclo[4.4.0]decane	0·267

TABLE 3.20

Shifts of trihalomethane stetching bands upon complex formation

Donor	$\dfrac{10^4 \, \Delta\nu_s}{\nu_s}$	Solvent (reference medium)	Ref.
	1. HCF$_3$		
dimethyl sulphoxide-d$_6$	92	CCl$_4$ [CCl$_4$]	[7]
	2. HCCl$_2$Br		
dimethyl sulphoxide	102	CCl$_4$ [CCl$_4$]	[6]
azine	142	CCl$_4$ [azine/CCl$_4$]	[6]
tetrabutylammonium chloride	395	CCl$_4$ [Bu$_4$NCl/CCl$_4$]	[6]
tetrabutylammonium bromide	346	CCl$_4$ [Bu$_4$NBr/CCl$_4$]	[6]
tetraheptylammonium iodide	302	CCl$_4$ [Hp$_4$NI/CCl$_4$]	[6]
	3. HCClBr$_2$		
dimethyl sulphoxide-d$_6$	142	CCl$_4$ [CCl$_4$]	[7]
	4. HCBr$_3$		
dimethyl sulphoxide	93	CCl$_4$ [DMSO/CCl$_4$]	[6]
dimethyl sulphoxide-d$_6$	165	CCl$_4$ [CCl$_4$]	[7]
azine	177	CCl$_4$ [azine/CCl$_4$]	[6]
azine-d$_5$	218	CCl$_4$ [azine/CCl$_4$]	[7]
triethylamine	596	C$_2$Cl$_4$ [C$_2$Cl$_4$]	[146]
tetrabutylammonium chloride	443	CCl$_4$ [Bu$_4$NCl/CCl$_4$]	[6]

TABLE 3.20–*continued*

Donor	$\dfrac{10^4 \; \Delta\nu_s}{\nu_s}$	Solvent (reference medium)	Ref.
tetrabutylammonium bromide	*399*	CCl_4 [Bu_4NBr/CCl_4]	[6]
tetraheptylammonium iodide	*346*	CCl_4 [Hp_4NI/CCl_4]	[6]
sodium iodide	*350*	$(CH_3)_2CO$ [$NaI/(CH_3)_2CO$]	[6]
	5. HCI_3		
dimethyl sulphoxide-d_6	133	CCl_4 [$DMSO/CCl_4$]	[7]
azine-d_5	183	CCl_4 [azine/CCl_4]	[7]
triethylamine	569	C_2Cl_4 [C_2Cl_4]	[146]

Values in italics are for the C–D stretching band of the deuterated acceptor molecule.

TABLE 3.21

^{1}H NMR Shifts of trihalomethanes upon complex formation

Initial medium	Chemical shift difference/ppm	Final medium	Ref.
	1. HCF_3		
cyclohexane	−1·0	benzene-d_6	[239]
cyclohexane	*0·69 ± 0·03*	oxolane *in cyclohexane*	[96]
cyclohexane	*1·480 ± 0·005* ($HC^{19}F_3$)	oxolane *in cyclohexane*	[95]
cyclohexane	*1·437 ± 0·007* ($DC^{19}F_3$)	oxolane *in cyclohexane*	[95]
	2. $HCBr_3$		
vapour	0·36$_2$	tetrachloromethane	[392]
cyclohexane	0·10	tetrachloromethane	[263]
cyclohexane	0·060	tetrachloromethane	[391]
vapour	0·45$_7$	carbon disulphide	[392]
vapour	0·52$_2$	iodomethane	[392]
vapour	0·67$_0$	diiodomethane	[392]
vapour	0·48$_2$	dibromomethane	[392]
vapour	0·34$_8$	dichloromethane	[392]
vapour	0·52$_3$	tribromomethane	[392]
cyclohexane	0·074	tribromomethane	[391]
vapour	0·35$_7$	trichloromethane	[392]
cyclohexane	0·08	trichloromethane	[263]
cyclohexane	0·076	trichloromethane	[391]
cyclohexane	−0·723	benzene	[391]
cyclohexane	−0·324	chlorobenzene	[391]
cyclohexane	−0·227	bromobenzene	[391]
cyclohexane	−0·537	thiole	[391]
cyclohexane	0·379	methanol	[391]
cyclohexane	0·451	ethanol	[391]

Values of complex shifts are indicated by italics.

was allowed to greet and acknowledge as equals. Such are the imprisonments imposed on its members by every culture of domination and submission, by its prescribed manner, vocabulary, tone of voice, gaze, and gesture—and the only explanation I can offer for the near-absence of blacks in this narrative of my family. In the documentary records, their names occasionally flash up and immediately disappear—in wills, for example, when ownership of a slave is designated for transmission from father to son, or in a memoir such as that of Sallie Ruberry Burgess, who mentions "Case," the buggy driver of Matthew Peterson Mayes, then promptly allows him to vanish. They were never people in a traditional Southern family; only properties, or props in social tableaux to which their subservience was crucial, yet always *presented* as purely incidental.

Among much else, I had hoped the taboos forbidding the recognition of people as something *more* than mere properties would be forgotten once I was out of the South for good. I imagined that life in large Northern cities would scrub away these inscriptions of class on my skin and in my gaze, and the powerful inclinations to ancestral roles inscribed on my mind. And for many years in my more officially egalitarian homeland, I half-believed this new life had accomplished just such a cleansing. Then came the moment in Aunt Vandalia's darkened, perfumed room when the eyes of her white servants turned toward me, quickening a feeling inside me I thought I would never again know, or have need of. It was mastery ancient and absolute, bringing with it feelings neither pleasant nor unpleasant—just odd, until I remembered that such had been the way of my father, grandfather, great-grandfather, on and on, back beyond the Civil War, the Revolution, even the English civil war. It was not an *American* stance, as we learned the word *American* in high school civics class. It was deeper than democracy, older than republicanism—the skill of my blood and kin and clan, the oldest of the old in what I was, conjured up in the house of my dead aunt.

The encounter was, in the end, faintly suffocating. I decided I needed something *to do*. So I had the curtained windows opened to light and breeze for the first time since Aunt Vandalia had entered upon her final dilapidation, and asked that dinner be brought to her desk, heaped high with unsorted papers, so I need not stop searching for her will and other papers, and looking for her burial instructions, if she had left any. Which, I decided late that night, she had not done. So I arranged the old lady's funeral according to the rites of Methodists, presided over the "visitation"—that gruesome obligatory last look at the deceased, caved-in face inflated with cotton and plastic injections, and made unspeakably livid with paint—and, at the funeral, did a reading from the Psalms.

Few people attended. My sister—co-inheritor with me of Aunt Vandalia's properties, a fact the servants and neighbors appeared not able to understand—decided not to fly down from Boston, where she lived. Most of Aunt Vandalia's friends and contemporary relatives were dead. One relation who wasn't—a middle-aged Texas kinsman of Uncle Alvin whom I could barely remember having seen before—kept sobbing about how Vandalia was just about his second mother. (This rather sloppy act was quickly halted after the funeral by the news that he'd inherited nothing, not even her car, which he openly coveted, after which he disappeared.) The service ended, a few mourners hobbled out to our family's burying ground, on a pleasant hillside easily visible from the kitchen in the days of cotton, now invisible from the house beyond densely reforested pastures.

There, beside her husband and her parents, Sister Erin and John Matthew Mays, near my father and other relations, and adjacent to the empty slot I had once accepted, then relinquished, I surrendered her to the ground—then turned to face the next duty on the long roster.

So far I had walked through these rites handily, as though following printed dance steps laid carefully on the ground in front of me; and I expected to continue doing so, right up until the moment I got on the plane back north and home. There was much paperwork needing attention, and a start on cleaning out the house to be made. I had spoken by telephone to my sister, who agreed that the disposal of the house and property should be done as quickly as legally possible. But regarding the house, I soon discovered, there was an unexpected problem to be resolved. It had to do with Jenny.

She had lived for fifty years on the far side of the south garden, in a plain, quaint bungalow my grandfather John had built and given to Aunt Vandalia when she married. Alvin and Vandalia were there only a few years before twin calamities fell on the newlyweds. One

cyclohexane	0·451	propanol-2	[391]
cyclohexane	0·437	2-methylpropanol-2	[391]
cyclohexane	0·472	diethyl ether	[391]
cyclohexane	0·639	diethyl ether	[296]
cyclohexane	0·514	dipropyl ether	[296]
cyclohexane	0·639	di(2-propyl) ether	[296]
cyclohexane	0·458	di(2-propyl) ether	[391]
cyclohexane	0·452	dibutyl ether	[296]
cyclohexane	0·520	1,4-dioxane	[296]
cyclohexane	0·785	oxolane	[296]
cyclohexane	*0·83 ± 0·02*	oxolane	
		in cyclohexane	[96]
cyclohexane	0·222	2-methylpropanal	[391]
cyclohexane	0·326	heptanal	[391]
cyclohexane	0·174	ethanoyl chloride	[391]
cyclohexane	0·622	propanone	[391]
vapour	$0·86_8$	propanone	[392]
hydrocarbon	0·50	propanone	[358]
cyclohexane	0·815	propanone	[296]
cyclohexane	0·828	heptanone-4	[296]
cyclohexane	0·636	heptanone-4	[391]
cyclohexane	0·782	2,6-dimethylheptanone-4	[296]
cyclohexane	0·615	cyclohexanone	[391]
cyclohexane	0·715	dimethyl methanamide	[391]
cyclohexane	0·97	dimethyl sulphoxide	[263]
vapour	$1·27_0$	dimethyl sulphoxide	[392]
vapour	$0·61_2$	ethanenitrile	[392]
cyclohexane	$0·21_5$	benzonitrile	[391]
cyclohexane	1·087	hexylamine	[391]
cyclohexane	1·139	cyclohexylamine	[391]
cyclohexane	1·210	2-methyl-2-propylamine	[391]
cyclohexane	0·978	di(2-propyl)amine	[391]
cyclohexane	0·871	triethylamine	[391]
cyclohexane	0·440	tripropylamine	[391]
cyclohexane	0·428	tributylamine	[391]
tetrachloromethane	*1·16 ± 0·17*	azine	
		in tetrachloromethane	[158]
tetrachloromethane	*2·29 ± 0·10*	tetrabutylammonium chloride	
		in tetrachloromethane	[158]
ethanenitrile	*2·34 ± 0·11*	tetrabutylammonium chloride	
		in ethanenitrile	[158]
tetrachloromethane	*1·82 ± 0·04*	tetrabutylammonium bromide	
		in tetrachloromethane	[158]
ethanenitrile	*1·57 ± 0·24*	tetrabutylammonium bromide	
		in ethanenitrile	[158]
tetrachloromethane	*1·06 ± 0·22*	tetraheptylammonium iodide	
		in tetrachloromethane	[158]
ethanenitrile	*0·83 ± 0·05*	tetrabutylammonium iodide	
		in ethanenitrile	[158]

3. HCl$_3$

vapour	$0·20_5$	tetrachloromethane	[392]
vapour	$0·31_2$	carbon disulphide	[392]
vapour	$0·26_7$	iodomethane	[392]
vapour	$0·37_7$	diiodomethane	[392]
vapour	$0·25_5$	dibromomethane	[392]
vapour	$0·15_0$	dichloromethane	[392]
vapour	$0·30_0$	tribromomethane	[392]
vapour	$0·18_8$	trichloromethane	[392]

was the abrupt disappearance of my grandfather's cotton and mercantile wealth at the outset of the Great Depression, and with it Alvin's job as overseer of a large ginning and farming operation. The other was my aunt's sudden blindness brought on by hemorrhages in her eyes, and, after a long recovery, her precariously fragile eyesight. The solution to their financial and health problems was to remove from their honeymoon cottage into the family home. The little house in the garden apparently stood vacant for some years, until it was rented out, in the late 1930s, to a Baptist minister and his wife, and their two daughters. One of these girls was Jenny.

And there she had dwelt, long after the marriage and departure of her sister and the deaths of her parents, until her working life as a secretary came to an end and she retired into solitude within the rented cottage. She was someone I had known about all my life, but, like so many other white people in the vicinity, not really known at all—a quiet white lady gradually aging in silence on the other side of the once-neat stands of tall tiger lilies, occasionally spoken of in my grandparents' house, but very rarely seen there.

I had not thought of her until one still evening just after Aunt Vandalia's burial, when I was sitting alone on the screened south verandah in the dark, wondering about the daunting work of clearing and cleaning and sorting and boxing that lay before us, and, in some vague way, perplexed—or perhaps just holding myself in perplexity as a remedy against the more serious work of mourning that, at this moment, I had no time to do. Jenny rapped gently at the porch's screen door, I invited her in, and our conversation began.

I cannot now reproduce or trace the twisting, indirect course our interchange took, any more than I could say exactly how the conversation with Billy and Lynda Dabbs, in Mayesville, worked its way out over the hours we spent together. Southern talk about things that matter is odd that way—convoluted, slow, and as winding as a bayou through bottom country, continually disappearing into irrelevancies, then reappearing at day's end, just in time for the resolution of the business that was probably inevitable all along. But Jenny's polite mission gradually became clear. It was to find out what I was going to do with her—or, as she put it, whether I intended to put her out. This matter was a worry, she explained, for two reasons. One was that she had nowhere to go, what with the high cost of Greenwood rents and all. The second was that, even if she found a place, the landlord might not let in her blind dogs.

Not a single Yankee known to me, and very few firmly urban Southerners, would understand fully the problem arising in this exchange. After all, the fate of Jenny and the blind dogs she cared for had never been my concern, and it had not become my responsibility—at least in any legal or otherwise binding sense—merely because I had inherited the house she'd rented from Aunt Vandalia. But the truth burning into the darkness on the verandah was that she *was* my concern, and that she knew it. Jenny was merely exercising a right that possessed no standing in law, nor in the modern history of contracts, but nevertheless had a certain secret, powerful position in a Southern family like mine. It is the right of a tenant to be taken care of, merely because she *is* a tenant. And in the end, when time came for the disposition of my aunt's Greenwood properties, the inevitable happened: Jenny was sold the cottage for roughly the cost of its demolition, and virtually given the grounds it stood on.

After the lady left me alone again on the verandah that mild, fragrant night, I found myself bemused by the quickness with which I had assumed a responsibility I had no obvious reason to take on. The feudal encounter with Jenny on the south porch would perhaps have proven more surprising had I not recently felt the intent gaze of Aunt Vandalia's servants on me, silently and subtly bestowing on me mastery unasked for. But I was gradually coming to recognize the quickening of those old powers in the blood I imagined had long before been emptied of urgency.

For no matter how much I wanted to return forever to the foreign, far-northern metropolis in which I'd found a life, and not think too much about the South again, Jenny's visit reminded me that the South remained reluctant to forget *me*—to let go of me, or perhaps anyone born on its terrain, nurtured by it, shaped in infancy by its heat and storms, dialect, and, above all, the codes and tactics of Southern existence handed down, father to son, through generations. Southerners without number who've left the South, I suspect, bear interesting memories of this common ground within, though they are usually unnoticed, and rarely summoned back into active imagination.

The longer one is away, of course, the quicker the large narrative

TABLE 3.21—*continued*

Initial medium	Chemical shift difference/ppm	Final medium	Ref.
vapour	0.46_7	propanone	[392]
hydrocarbon	0·15	propanone	[358]
vapour	0.57_3	dimethyl sulphoxide	[392]
cyclohexane	0.43 ± 0.03	oxolane	
		in cyclohexane	[96]
vapour	0.32_8	ethanenitrile	[392]
tetrachloromethane	0.38 ± 0.06	azine	
		in tetrachloromethane	[158]
ethanenitrile	-0.31 ± 0.01	tetrabutylammonium chloride	
		in ethanenitrile	[158]
ethanentrile	-0.32 ± 0.01	tetrabutylammonium bromide	
		in ethanenitrile	[158]
ethanenitrile	-0.36 ± 0.01	tetrabutylammonium iodide	
		in ethanenitrile	[158]
	4. $HC(NO_2)_3$		
cyclohexane	-1.855	benzene	
		in cyclohexane	[179]
cyclohexane	-1.880	methylbenzene	
		in cyclohexane	[179]
cyclohexane	-1.950	1,2-dimethylbenzene	
		in cyclohexane	[179]
cyclohexane	-1.897	1,3-dimethylbenzene	
		in cyclohexane	[179]
cyclohexane	-1.987	1,4-dimethylbenzene	
		in cyclohexane	[179]
cyclohexane	-1.997	1,3,5-trimethylbenzene	
		in cyclohexane	[179]
cyclohexane	-2.145	1,2,4,5-tetramethylbenzene	
		in cyclohexane	[179]
cyclohexane	-2.024	pentamethylbenzene	
		in cyclohexane	[179]
cyclohexane	-2.383	hexamethylbenzene	
		in cyclohexane	[179]

TABLE 3.22

Vibrational frequency shifts of mono- and dihalo substituted HCXYZ compounds upon complex formation in CCl_4 solution [7]

X	Y	Z	$10^4 \, \Delta\nu_s/\nu_s$	
			with dimethyl sulphoxide-d_6	with azine-d_5
Br	Br	CN	266	377
Br	Br	CBr_3	193	290
Br	Br	$CHBr_2$	137	
Cl	Cl	CN	220	308
Cl	Cl	CCl_3	137	214
Cl	Cl	$CHCl_2$	83	

Cl	Cl	CCl_2CCl_3	149	216
Cl	Cl	$COOCH_3$	86	
Cl	CCl_3	CCl_3		277
Cl		$CCl_2\overset{\diagdown\diagup}{-}CCl_2$		289

TABLE 3.23

Formation constants and enthalpies of formation for complexes of
1-hydroperfluoroheptane [8]

Donor	T/K	$K_\phi/\text{mol mol}^{-1}$	$\Delta_\phi H/\text{kJ mol}^{-1}$
propanone	253	4.0 ± 1.0	-10.5 ± 2.1
	303	2.0 ± 0.6	
	353	1.0 ± 0.4	
pentanone-3	253	3.8 ± 0.8	
	303	2.2 ± 0.6	-7.5 ± 1.7
	353	1.4 ± 0.4	
diethyl ether	253	1.0 ± 0.4	
	303	0.6 ± 0.2	-7.1 ± 1.7
	353	0.4 ± 0.2	
triethylamine	253	7.0 ± 2.0	
	303	1.5 ± 0.5	-20.5 ± 3.8
	353	0.45 ± 0.15	

TABLE 3.24

Enthalpies of mixing of HCXYZ with electron donor molecules

Donor	T/K	$\Delta_m H/\text{kJ mol}^{-1}$	Ref.
	1. $HCCl_2(C_6H_5)$		
dimethyl ethanamide	276	-2090	[259]
diethyl ether	276	-970	[259]
propanone	276	-835	[259]
	2. $HCCl(C_6H_5)_2$		
dimethyl ethanamide	287	-870	[259]
diethyl ether	287	0	[259]
	3. $HC(OC_2H_5)_3$		
ethyl ethanoate	276	0	[259]
	4. $HCF_2(C_6F_{13})$		
propanone	273	-920^a	[9]
	308	-710^a	[9]
	5. $HCCl_2CHCl_2$		
ethyl ethanoate	276	-2545	[259]
propanone	276	-2550	[259]
benzene	278	-525	[259]
nitrobenzene	278	-585	[259]
benzenecarbonitrile	276	-1515	[259]
azine	276	-2530	[259]

[a] Mole fraction of propanone $= 1/3$.

TABLE 3.25

^{1}H NMR Shifts upon formation of HCXYZ complexes

Initial medium	Chemical shift difference/ppm	Final medium	Ref.
		1. $HCCl_2CN$	
heptane	0·78	1,4-dioxane	
		in heptane	[193]
heptane	−1·83	benzene	
		in heptane	[193]
heptane	−1·87	methylbenzene	
		in heptane	[193]
heptane	−1·93	1,3,5-trimethylbenzene	
		in heptane	[193]
heptane	−2·10	pentamethylbenzene	
		in heptane	[193]
heptane	−2·32	1-methylnaphthalene	
		in heptane	[193]
		2. $HCCl_2Br$	
tetrachloromethane	0·09$_5$	carbon disulphide	[392]
tetrachloromethane	0·19$_7$	iodomethane	[392]
tetrachloromethane	0·35$_8$	diiodomethane	[392]
tetrachloromethane	0·16$_0$	dibromomethane	[392]
tetrachloromethane	−0·00$_2$	dichloromethane	[392]
tetrachloromethane	0·17$_8$	tribromomethane	[392]
tetrachloromethane	0·01$_5$	trichloromethane	[392]
tetrachloromethane	0·52$_7$	propanone	[392]
tetrachloromethane	1·05$_0$	dimethyl sulphoxide	[392]
tetrachloromethane	−0·19$_5$	ethanenitrile	[392]
		3. $HCCl_2CH_2Cl$	
cyclohexane	0·12	tetrachloromethane	[263]
cyclohexane	0·23	1,1,2-trichloroethane	[263]
cyclohexane	0·95	dimethyl sulphoxide	[263]
		4. $HCCl_2CHCl_2$	
cyclohexane	0·14	tetrachloromethane	[263]
cyclohexane	0·22	1,1,2,2-tetrachloroethane	[263]
cyclohexane	1·16	dimethyl sulphoxide	[263]
		5. $HCCl_2CCl_3$	
cyclohexane	0·14	tetrachloromethane	[263]
cyclohexane	0·14	pentachloroethane	[263]
cyclohexane	1·52	dimethyl sulphoxide	[263]
		6. $HCBr_2CH_2Br$	
cyclohexane	0·10	tetrachloromethane	[263]
cyclohexane	0·22	1,1,2-tribromoethane	[263]
cyclohexane	0·85	dimethyl sulphoxide	[263]
		7. $HCBr_2CHBr_2$	
cyclohexane	0·12	tetrachloromethane	[263]
cyclohexane	0·20	1,1,2,2-tetrabromoethane	[263]
cyclohexane	0·97	dimethyl sulphoxide	[263]
		8. $HCBr_2CN$	
cyclohexane	0·12	tetrachloromethane	[263]
cyclohexane	0·22	dibromoethanenitrile	[263]
cyclohexane	1·27	dimethyl sulphoxide	[263]

of agrarian Southern existence disappears from conscious life, leaving only harmless traces and artifacts and folkways to remind us of the complex matrix of values in which we were formed. But these traces remain. Though the ingredients for it are a bit hard to find in Toronto, for example, the cuisine of New Year's Day in my household, dawn to midnight, consists only of black-eyed peas, plain cornbread, and, if we can get them, collard greens. These were the traditional staples of the South's poorest whites and blacks, and, to this day, are believed by superstitious Southerners (like me, and like Margaret, who comes from Arkansas) to bring prosperity in the coming twelve months.

Then there's the way I talk, and write. From time to time, a bewildered editor at my newspaper will call me over after stumbling on a peculiar usage or rhetorical turn in a story I've written—something the editor's never before seen, but that had seeped up again into my journalistic language from the pool of regional language in the bottom of my mind. When Margaret and I are alone together, or with our daughter Erin—a Canadian born and bred, always ready to smile knowingly when her parents' Canadianized speech begins to slow down to Southern speed—we often slip into the kitchen dialect of our homeland, using the *ain'ts* and *y'alls* that we never allow ourselves in public. But both of us put a certain store by good table manners and well-kept decorum, and something we call "fitting conduct"—hard to define exactly, but easy to recognize in people who are conducting themselves unfittingly—along with other trivial values and ideas still alive in us, though formed in institutions of Southern agrarian hierarchy now vanished or dying.

What I am talking about, of course, is not being Southern—at least in the sense I have used it in most of this book, with its connotations of grounding in a place and within some sustained pattern of life. These matters of manner and diet and such together comprise Southern culture in the most narrow sense: the heap of furniture, architectural style, peculiar notions, folkways, and customs—grits and fiery Baptist oratory and Mississippi Delta and good ol' boys with beer bellies as well as a well-laid table and a respect for elders—that have survived the gradual subsidence of Southern civilization's sustaining traditions. Even in the contemporary South, these remnants are increasingly meaningless, detached signifiers and souvenirs left floating on the surface after our Southern Atlantis sank under the tidal waves of military defeat, and industrial urbanization.

There is no clearer testimony to this general vanishing than the recent distillation of Southern memories into an academic discipline. The University of Mississippi's Center for the Study of Southern Culture, and the center's vast and comprehensive *Encyclopedia of Southern Culture*, published in 1989, both stand as monuments to the resistance of Southern scholars to let certain artifacts, social quirks, and distinctive regional tastes disappear from memory altogether. But however effective the Ole Miss institute, however entertaining and useful its *Encyclopedia*, both are symptoms of a large extinction, much like museums devoted to the preservation of Minoan epigraphy or Babylonian idols and tablets. Were Southern civilization alive and contentious, like the urban black ghettos of the South blooming with both crime and astonishing musicians and artists, it would be invisible to scholars as a topic of celebration, *because* it was alive, hence subversive to the grand historical narratives of the Frontier and the Great Immigrations by which Americans explain America to themselves.

Yet, at least in some of us, the binding rules and tales of Southern civilization seem to persist at the deep levels of consciousness where the anthropological oddments of "Southern culture," in the academic sense, are irrelevant, even unknown. (Among the many delights I found while reading through the huge *Encyclopedia of Southern Culture* were things, people, "traditions" that were presumably very Southern, of which I'd never heard.) But in exchanges such as the one with my elderly tenant on the south verandah, old and largely forgotten ideas surface from the deep agrarian civilization in the South that came before now, and even before the Civil War, and that had persisted even through my immediate ancestors' abandonment of the countryside and its ways, and their move into the cities and the professions.

The central idea, as always, is imbedded in a certain understanding of the land. The property left me by my aunt was not, in the end, merely an abstraction, some geometrical diagram laid out on a surveyor's map. It was a complex event, a *situation* in an older sense of that word—close to *site*, though without that word's overtones of empty Euclidean infinity. The situation in Greenwood was composed

9. $HCBr(CH_3)CH_2Br$

cyclohexane	0·07	tetrachloromethane	[263]
cyclohexane	0·13	1,2-dibromopropane	[263]
cyclohexane	0·47	dimethyl sulphoxide	[263]

TABLE 3.26

Formation constants and NMR complex shifts of halide ion complexes of β-tetraacetylglucopyranosides in CH_3CN solvent at 308 K [258] [Counter ion = $(C_2H_5)_4N^+$]

β-Substituent	Donor ion	K_ϕ/dm^3 mol^{-1}	Δ_c/ppm				
			1	3	5	*ortho*	*meta*
	Cl$^-$	1·07 ± 0·06	1·52 ± 0·06	0·35 ± 0·07	1·21 ± 0·12		
	Cl$^-$	0·67 ± 0·15	1·14 ± 0·22	0·26 ± 0·08	0·99 ± 0·21		
$C_6H_4OCH_3$-4	Cl$^-$	0·13 ± 0·06	2·8 ± 1·4	0·5 ± 0·5	3·2 ± 1·8	0·6 ± 0·3	0·2 ± 0·1
C_6H_5	Cl$^-$	0·64 ± 0·05	1·29 ± 0·10	0·22 ± 0·06	1·04 ± 0·13	0·50 ± 0·11	0·00 ± 0·03
$C_6H_4NO_2$-4	Cl$^-$	2·03 ± 0·11	1·31 ± 0·04	0·15 ± 0·01	1·13 ± 0·08	0·37 ± 0·04	
$C_6H_4NO_2$-4	Br$^-$	1·38 ± 0·12	1·23 ± 0·06	0·08 ± 0·04	1·22 ± 0·21		
$C_6H_4NO_2$-4	I$^-$	1·05 ± 0·14	0·78 ± 0·08	0·00 ± 0·05	1·07 ± 0·27		

TABLE 3.27

Solubility of dichloromethane in donor solvents at 305·4 K [88]

Solvent	Solubility^a/mol mol^{-1}
phenol	0·130
phenyl methyl ether	0·328
1,4-dimethoxycyclohexane	0·442
3,6,9-trioxaundecane	0·520
2,5,8,11,14-pentaoxapentadecane	0·602
ethanoic acid	0·174
propanoic acid	0·218
ethyl butan-3-one-1-oate	0·410
ethyl 2,2-diethylbutan-3-one-1-oate	0·468
3,6,9-trioxaundecan-2-one	0·509
triphenoxyphosphine	0·412
tris(methoxyphenoxy)phosphine oxide	0·470
triethoxyphosphine oxide	0·541
tripropoxyphosphine oxide	0·567
tributoxyphosphine oxide	0·574
cyclohexyl amine	0·377
methanamide	0·038
methyl methanamide	0·202
dimethyl methanamide	0·261
methyl ethanamide	0·269
dimethyl ethanamide	0·452
methyl cyclohexyl ethanamide	0·521
ethyl cyclohexyl ethanamide	0·530
methyl cyclohexyl butanesulphonamide	0·490

TABLE 3.27—*continued*

Solvent	Solubility^a/mol mol^{-1}
butanone oxime	0·244
benzaldehyde	0·349
2-hydroxybenzaldehyde	0·315
heptanal	0·388
2,4,6-trimethyl-1,3,5-trioxane	0·374
cyclohexanone	0·421
cyclohexenone-3	0·349
pentanedione-2,4	0·394
hexanedione-2,5	0·434
1,3,5-trimethylbenzene	0·317^b
pentanenitrile	0·452^b
octanenitrile	0·405^c
benzenecarbonitrile	0·359^c
decanedinitrile	0·435^c
hexanedinitrile	0·316^c
pentanedinitrile	0·261^c
butanedinitrile	0·158^c

^{a} The value expected on the basis of Raoult's Law is 0·311 mol mol^{-1}.
^{b} Ref. [89].
^{c} Ref. [90].

TABLE 3.28

Enthalpies of mixing of H_2CX_2 [259]

Donor	T/K	$\Delta_m H$/kJ mol^{-1}
	1. H_2CCl_2	
diethyl ether	276	−1025
dimethyl ethanamide	276	−2205
	2. H_2CBr_2	
diethyl ether	276	−835
dimethyl ethanamide	276	−1675
	3. H_2CI_2	
diethyl ether	280	0
dimethyl ethanamide	280	−420
	4. $H_2C(OC_2H_5)_2$	
dimethyl ethanamide	276	0

TABLE 3.29

Formation constants of H_2CX_2 complexes

Donor	T/K	K	Units	Solvent	Ref.
	1. H_2CCl_2				
benzene	313	0·12	M^{-1}	C_7H_{16}	[193]
methylbenzene	313	0·15	M^{-1}	C_7H_{16}	[193]
1,3,5-trimethylbenzene	313	0·17	M^{-1}	C_7H_{16}	[193]

1-methylnaphthalene	313	0·22	M^{-1}	C$_7$H$_{16}$	[193]
tris(dimethylamino)phosphine oxide	293	1·4	M^{-1}	c-C$_6$H$_{12}$	[291]
1,4-dioxane	313	0·20	M^{-1}	C$_7$H$_{16}$	[193]
dioctyl ether	303·15	0·280 ± 0·010	M^{-1}	–	[352]
	313·15	0·260 ± 0·010	M^{-1}	–	[352]
	323·15	0·241 ± 0·010	M^{-1}	–	[352]
	333·15	0·223 ± 0·010	M^{-1}	–	[352]
dioctyl sulphide	303·15	0·359 ± 0·010	M^{-1}	–	[352]
	313·15	0·335 ± 0·010	M^{-1}	–	[352]
	323·15	0·313 ± 0·010	M^{-1}	–	[352]
	333·15	0·294 ± 0·010	M^{-1}	–	[352]
2. H$_2$CClBr					
dioctyl ether	303·15	0·290 ± 0·010	M^{-1}	–	[352]
	313·15	0·264 ± 0·010	M^{-1}	–	[352]
	323·15	0·241 ± 0·010	M^{-1}	–	[352]
	333·15	0·218 ± 0·010	M^{-1}	–	[352]
dioctyl sulphide	303·15	0·407 ± 0·010	M^{-1}	–	[352]
	313·15	0·377 ± 0·010	M^{-1}	–	[352]
	323·15	0·350 ± 0·010	M^{-1}	–	[352]
	333·15	0·326 ± 0·010	M^{-1}	–	[352]
3. H$_2$CBr$_2$					
dioctyl ether	303·15	0·297 ± 0·010	M^{-1}	–	[352]
	313·15	0·272 ± 0·010	M^{-1}	–	[352]
	323·15	0·249 ± 0·010	M^{-1}	–	[352]
	333·15	0·227 ± 0·010	M^{-1}	–	[352]
dioctyl sulphide	303·15	0·455 ± 0·010	M^{-1}	–	[352]
	313·15	0·424 ± 0·010	M^{-1}	–	[352]
	323·15	0·396 ± 0·010	M^{-1}	–	[352]
	333·15	0·370 ± 0·010	M^{-1}	–	[352]

TABLE 3.30

Thermodynamic parameters for H$_2$CX$_2$ complexes

Donor	$\Delta_\phi H^\circ$/kJ mol^{-1}	$\Delta_\phi S^\circ$/J K^{-1} mol^{-1}	Solvent	Ref.
1. H$_2$CCl$_2$				
benzene	−7·1	−40·6	C$_7$H$_{16}$	[193]
methylbenzene	−8·4	−41·8	C$_7$H$_{16}$	[193]
1,3,5-trimethylbenzene	−8·8	−43·5	C$_7$H$_{16}$	[193]
1-methylnaphthalene	−5·9	−31·0	C$_7$H$_{16}$	[193]
1,4-dioxane	−5·4	−31·0	C$_7$H$_{16}$	[193]
dioctyl ether	−6·3$_6$	−31·5	–	[352]
dioctyl sulphide	−5·6$_1$	−27·0	–	[352]
2. H$_2$CClBr				
dioctyl ether	−7·9$_5$	−36·5	–	[352]
dioctyl sulphide	−6·2$_3$	−27·9	–	[352]
3. H$_2$CBr$_2$				
dioctyl ether	−7·4$_9$	−34·8	–	[352]
dioctyl sulphide	−5·7$_7$	−25·6	–	[352]

of dirt and architecture, rose gardens and chicken houses and the ruins of Uncle Alvin's rabbit hutches. But also involved was history—the memories and the facts of lives lived, the sum of traces left those who had dwelt there, all of them, enduring or fading or coming forward through time into living memory. Jenny's long tenancy in Aunt Vandalia's cottage, her sorrow at the deaths of Sister Erin and John Matthew Mays, the joy she got from tending the blind dogs she'd gathered up from heaven knows where, were among such memories inscribed on the land, hence inheritable.

In my mind then and now, this idea stands as a sound understanding of property, and surely not something to be lightly disregarded; even though, following in the path of the great-grandfather who left South Carolina for a new life, I had myself disregarded it for a long time. But those memories about the Southern earth were as real to me at that moment with Jenny as the frame house raised by my grandparents, as the magnificent pecan tree in the south garden and the beautiful old roses on the house's north wall, continuing to bloom sweetly and abundantly on clamoring vines each summer for nobody, once everyone inside the house had become too old or ill to tend them.

My final visit to the house built in the northwest Louisiana town of Greenwood by John Matthew Mays and Sister Erin, my grandparents, lived in by Aunt Vandalia and Uncle Alvin, and, briefly, by me, took place on a hot morning after Jenny's visit, when the old dwelling's emptying was done. The labor of sorting strata of papers and possessions and furniture, stacked eighty years deep, was finished. Packed up, all the legalities and other abstractions that attend dying seen to, I was ready to head north, for the first leg of the long trip home. My cousin Lane had moved in to look after the place until it could be disposed of—though, that morning, he was out, and I was alone in the near-empty rooms. Nothing remained to be done, except the performance of last rites.

There was no formula, no printed ritual—merely walking and pausing, and a certain remembering at each station along the way. The route of this farewell and release took me through the back gallery, with its memories of hot biscuits served at breakfast by my grandmother's cook Peggy, and through the kitchen and family bed-

TABLE 3.31

Solvent effects on the C–Cl stetching frequencies of dichloromethane
[referred to vapour–phase H_2CCl_2] [162, 290]

Solvent	$10^4 \, \Delta\nu/\nu$	
	ν_3	ν_9
hexane	184	
cyclohexane	198	200
tetrachloromethane		228
carbon disulphide	244	235
tetrachloroethene		228
benzene	283	
diethyl ether	283	297
1,4-dioxane	329	317
ethyl ethanoate	303	311
propanone	323	324
nitromethane	303	311
ethanenitrile	316	311
triethylamine		297

TABLE 3.32

^{1}H NMR Solvent shifts and complex shifts of H_2CX_2 upon complex formation

Initial medium	Chemical shift difference/ppm	Final medium	Ref.
	1. H_2CCl_2		
vapour	0.50_2	tetrachloromethane	[392]
cyclohexane	0.108	tetrachloromethane	[391]
cyclohexane	0.152	tetrachloromethane	[263]
vapour	0.59_7	carbon disulphide	[392]
heptane	-1.40	benzene *in heptane*	[193]
heptane	-1.35	methylbenzene *in heptane*	[193]
heptane	-1.37	1,3,5-trimethylbenzene *in heptane*	[193]
heptane	-1.73	1-methylnaphthalene *in heptane*	[193]
heptane	0.47	1,4-dioxane *in heptane*	[193]
vapour	0.64_8	iodomethane	[392]
vapour	0.90_5	diiodomethane	[392]
vapour	0.65_0	dibromomethane	[392]
vapour	0.49_0	dichloromethane	[392]
cyclohexane	0.147	dichloromethane	[391]
cyclohexane	0.198	dichloromethane	[263]
vapour	0.72_8	tribromomethane	[392]
vapour	0.50_5	trichloromethane	[392]
vapour	0.66_2	ethanenitrile	[392]
cyclohexane	0.52	dimethyl methanamide	[391]
vapour	0.64_7	propanone	[392]
vapour	1.03_3	dimethyl sulphoxide	[392]

Values given in italics are complex shifts.

rooms, and, at last, to the part I had always found grandest: the more public, sociable sequence of south verandah, living room, study, and formal dining room, each section separated from its neighbors by double doors, shut in winter or when privacy was wanted, opened for entertaining, and for breeze in summer.

As a small child, on visits to this house of rhyming rooms, I had listened to the men talking politics and cotton under the ceiling fan on the south porch, and played on the living room carpet with things from Sister Erin's cabinet of wonders—a bisque figurine named Mr. Penguin, seashells and delicate glass animals from Mexico, fossils and arrowheads, and books with odd pictures in them I later realized were mesmeric, phrenological, astrological. (My grandmother had been inclined to an interest in such matters, for a while.) When I was a little older and ready for them, the ten black volumes of *Our Wonder World*, the green and bark-brown books of forest lore and woodcraft by Ernest Thompson Seton, huge Buster Brown comic books and other delights that had belonged to my father and aunt would be brought from their shelves behind glass doors. And among those books and toys, I was set for the day—happily marooned on this vast island of carpet, round which the household business went on as usual.

The last farewells were said in the attic loft where my father had slept as a boy. Looking out the high windows that lit John Bass's garret, I saw much of the old town of my grandfather, and remembered the broad peaceful pasture that once spread almost up to the back garden, dew or frost gleaming on long grass and mock-orange bushes, cows drifting like clouds against a sky of deep summer green. Those visions had vanished with the passage of years, as the pasture had been transformed into a trailer park, and once-familiar sights had been concealed by trees and brush. Since I was a child, the distance of Greenwood from the world beyond had lessened with the coming of better roads, then collapsed altogether in the 1960s, when the interstate highway was rammed through the red-clay hills nearby. What I remembered was gone—the pacific unity of the old town and the old house, the settled social hierarchies that had always seemed of a piece with the pastures falling away down the hill from the village, the books in John and Sister Erin's library, the scents of the house and the fragrance of the air on summer evenings.

At last the time had come to say good-bye, not only to the house, but to Greenwood, as it persisted in memory and desire.

Beyond my father's high windows, and a patch of pasturage, just up the road from our front gate, stood cousin Miller Mays's house, put up on the site of the town's first building, in 1836—a trading post for the Caddoan forest people, hunters of bear and turkey in that hilly wilderness above the Red River valley, until forced on and away west. The planters and settlers and merchants came later, upriver and overland from the more easterly states of the South, carving from the forest a town square, in expectation of civic importance that was not to materialize, and gathering their large houses around it. I could just see the square from my father's window—an expanse of weed and a barren playground now, its great old oaks ripped down by a tornado years before, worthless yet absurdly protected by an ugly steel mesh fence.

As I looked down at the square and the fading town, remembrances of matters I could not have witnessed came back to me, from some obscure corner in the mind, where they had been put years before by reading or telling—memories of duels fought under the town square's two great oaks; of music in the ballrooms of the antebellum Cresswell and Parson Doty and Whitworth houses, of the notorious flight of Miss Lucia Moore, in the 1870s, to New York and a scandalous life on the stage, of names entwined with my own—Caleb Eubanks, Dr. Flournoy, Miss Adeline Fortson, the Trospers and Brysons and McClurgs. To the south lay a still-mighty stand of trees called simply The Grove, where troopers marching home from the War of 1812 were said to have camped for a night. Yet could not the great trees themselves have been turned into soldiers by the mutating force of storytelling? What do we dare believe, or disbelieve, when the land awakens thoughts of dwelling?

Down the hill from the house, too, was what had once been the lot on which my grandfather built his first cotton gin in 1910, to serve the farmers who came in wagons heaped high with snowy bolls, iron wheels rattling up from the deep country on dirt roads—a site now buried underneath the gray, noisy ribbons of expressway. Yet how real was that peaceable vision? Had it ever existed? Or did this Southern land wish it had existed, and whispered its dream into my mind as I stood at my father's windows?

Thinking these things, I noticed one thing that had not changed: the sky, as hot, still, and intensely blue as I remembered it as a boy, flecked with the incandescent clouds—the sky into which the tall boy my father had been careened in his first little airplane, disappearing into the sky, then reappearing above The Grove, the tall old houses, the architecture of memory the town was and is. Or merely *was*; for, like so many other Southern towns that had not merely declined into nothingness, Greenwood had been absorbed by a nearby city, eradicating old landmarks, transfiguring and erasing the old places I had known as a child. All that remained was the *idea* of this Southern town—a far more compelling idea than the Old South of popular imagination, with its great columned porticos and odiferous magnolias. Gone, as well, was any reason for me *not* to say the last farewell to the old house in Greenwood, turn the large iron key in the door lock for the last time, and leave.

Not long after I did so, I began the journeys to my ancestral Southern homelands chronicled in this book. I did not travel alone.

No one—finding himself suddenly estranged from the familiar, then deciding to traverse the forest fastnesses of memory in an attempt to retrieve his truest self—ever does. Before striking out, one never knows exactly where this help will come from; but at the first step beyond the known it appears. For Dante, the guide given him in the dark wood was Virgil—the personification of Reason and Antiquity, the old commentaries on the *Divina Commedia* tell us. For the knights of the Round Table in search of the Holy Grail, the guides are many—monks and spectral maidens and mysterious helmeted paladins appearing to give advice on the next crucial step, then disappearing again into the darkness from which they came. But this appearance of the Guide is not merely a literary trope, however venerable and commonplace in our Western literatures. I would argue that it is simply everybody's everyday experience, given beautiful and comprehensible shape by the geniuses of our tradition. (Take off the horns and hides, someone has said of the characters in Richard Wagner's magnificent music dramas, and what you have are people with the same problems as you and me, and the same difficulty working them out. The same could be said of Homer's gods and heroes, or Dostoyevsky's antiheroes, or the nonhero Leopold Bloom,

cyclohexane	0·59	dimethyl sulphoxide	[263]
cyclohexane	*0·95*	tris(dimethylamino)phosphine oxide *in cyclohexane*	[291]

2. H_2CClCN

tetrachloromethane	0·11$_2$	carbon disulphide	[392]
tetrachloromethane	0·22$_5$	iodomethane	[392]
tetrachloromethane	0·50$_7$	diiodomethane	[392]
tetrachloromethane	0·24$_8$	dibromomethane	[392]
tetrachloromethane	−0·00$_7$	dichloromethane	[392]
tetrachloromethane	0·07	chloroethanenitrile	[391]
tetrachloromethane	0·27$_8$	tribromomethane	[392]
tetrachloromethane	0·01$_3$	trichloromethane	[392]
tetrachloromethane	0·15$_7$	ethanenitrile	[392]
tetrachloromethane	0·56	dimethyl methanamide	[391]
tetrachloromethane	0·29$_7$	propanone	[392]
tetrachloromethane	0·65$_8$	dimethyl sulphoxide	[392]

3. H_2CClBr

vapour	0·49$_0$	tetrachloromethane	[392]
cyclohexane	0·107	tetrachloromethane	[391]
cyclohexane	0·133	tetrachloromethane	[263]
vapour	0·59$_2$	carbon disulphide	[392]
vapour	0·63$_8$	iodomethane	[392]
vapour	0·90$_8$	diiodomethane	[392]
vapour	0·64$_7$	dibromomethane	[392]
cyclohexane	0·160	chlorobromomethane	[391]
cyclohexane	0·21	chlorobromomethane	[263]
vapour	0·45$_5$	dichloromethane	[392]
vapour	0·69$_0$	tribromomethane	[392]
vapour	0·49$_0$	trichloromethane	[392]
vapour	0·63$_0$	ethanenitrile	[392]
cyclohexane	0·51	dimethyl methanamide	[391]
vapour	0·64$_0$	propanone	[392]
vapour	1·00$_7$	dimethyl sulphoxide	[392]
cyclohexane	0·51	dimethyl sulphoxide	[263]

4. H_2CBr_2

vapour	0·49$_7$	tetrachloromethane	[392]
cyclohexane	0·15	tetrachloromethane	[263]
cyclohexane	0·08	tetrachloromethane	[391]
vapour	0·58$_8$	carbon disulphide	[392]
vapour	0·63$_0$	iodomethane	[392]
vapour	0·87$_0$	diiodomethane	[392]
vapour	0·64$_2$	dibromomethane	[392]
cyclohexane	0·153	dibromomethane	[391]
cyclohexane	0·20	dibromomethane	[263]
vapour	0·46$_0$	dichloromethane	[392]
vapour	0·68$_8$	tribromomethane	[392]
vapour	0·48$_0$	trichloromethane	[392]
vapour	0·64$_0$	ethanenitrile	[392]
cyclohexane	0·50	dimethyl methanamide	[391]
vapour	0·65$_2$	propanone	[392]
vapour	1·06$_7$	dimethyl sulphoxide	[392]
cyclohexane	0·60	dimethyl sulphoxide	[263]

5. H_2CBrI

cyclohexane	0·07	tetrachloromethane	[391]
cyclohexane	0·11	bromoiodomethane	[391]
cyclohexane	0·38	dimethyl methanamide	[391]

TABLE 3.32–*continued*

Initial medium	Chemical shift difference/ppm	Final medium	Ref.
		6. H_2Cl_2	
vapour	0.34_3	tetrachloromethane	[392]
cyclohexane	0.051	tetrachloromethane	[391]
vapour	0.47_5	carbon disulphide	[392]
vapour	0.43_8	iodomethane	[392]
vapour	0.67_0	diiodomethane	[392]
cyclohexane	0.066	diiodomethane	[391]
vapour	0.45_2	dibromomethane	[392]
vapour	0.29_3	dichloromethane	[392]
vapour	0.51_5	tribromomethane	[392]
vapour	0.32_3	trichloromethane	[392]
vapour	0.42_2	ethanenitrile	[392]
cyclohexane	0.26	dimethyl methanamide	[391]
vapour	0.40_2	propanone	[392]
vapour	0.50_5	dimethyl sulphoxide	[392]
		7. H_2CClCH_3	
cyclohexane	0.10	tetrachloromethane	[263]
cyclohexane	0.05	chloroethane	[263]
cyclohexane	0.25	dimethyl sulphoxide	[263]
		8. H_2CClCH_2Cl	
cyclohexane	0.12	tetrachloromethane	[263]
cyclohexane	0.19	1,2-dichloroethane	[263]
cyclohexane	0.32	dimethyl sulphoxide	[263]
		9. $H_2CClCHCl_2$	
cyclohexane	0.11	tetrachloromethane	[263]
cyclohexane	0.16	1,1,2-trichloroethane	[263]
cyclohexane	0.40	dimethyl sulphoxide	[263]
		10. $H_2CClCCl_3$	
cyclohexane	0.11	tetrachloromethane	[263]
cyclohexane	0.15	1,1,1,2-tetrachloroethane	[263]
cyclohexane	0.63	dimethyl sulphoxide	[263]
		11. H_2CBrCH_3	
cyclohexane	0.15	tetrachloromethane	[263]
cyclohexane	0.15	bromoethane	[263]
cyclohexane	0.25	dimethyl sulphoxide	[263]
		12. $H_2CBrCH_2CH_2Br$	
cyclohexane	0.09	tetrachloromethane	[263]
cyclohexane	0.10	1,3-dibromopropane	[263]
cyclohexane	0.18	dimethyl sulphoxide	[263]
		13. $H_2CBrCHBrCH_3$	
cyclohexane	0.15	tetrachloromethane	[263]
cyclohexane	0.15	1,2-dibromopropane	[263]
cyclohexane	0.26	dimethyl sulphoxide	[263]
		14. $H_2CBrCHBr_2$	
cyclohexane	0.11	tetrachloromethane	[263]
cyclohexane	0.18	1,1,2-tribromoethane	[263]
cyclohexane	0.35	dimethyl sulphoxide	[263]

Values given in italics are complex shifts.

TABLE 3.33

^{1}H NMR Solvent shifts of 3-X-propyne-1

Solvent	Δ_s/ppm^a	
	X = Cl^b	X = Br^c
(pure compounds)	0·26	0·70
heptane		−0·09
2,2,4-trimethylpentane		−0·04
tetrachloromethane		0·39
carbon disulphide		0·52
dichloromethane		0·52
1,2-dichloroethane		0·60
dibromomethane		1·13
nitrobenzene	0·34	
nitromethane	0·31	
ethanenitrile	0·30	
1,4-dioxane	0·26	
propanone	0·36	0·03
dimethyl methanamide	0·52	
thiole	−0·25	
fluorobenzene	−0·13	
benzene	−0·45	−1·02
methylbenzene	−0·49	

^{a} Reference solvent is cyclohexane.
^{b} Ref. [171]; ± 0·03 ppm.
^{c} Ref. [399]; T = 294 K.

TABLE 3.34

^{1}H NMR Gas-to-solution shifts of H_3CX [392]

Solvent	Chemical shift difference/ppm			
	X = Cl	X = Br	X = I	X = CN
tetrachloromethane	0·46$_3$	0·50$_2$	0·48$_8$	0·56$_7$
carbon disulphide	0·56$_0$	0·60$_5$	0·61$_5$	0·65$_7$
iodomethane	0·53$_3$	0·57$_5$	0·56$_3$	0·69$_8$
diiodomethane	0·86$_8$	0·90$_8$	0·88$_2$	1·03$_7$
dibromomethane	0·58$_3$	0·62$_7$	0·60$_8$	0·71$_8$
dichloromethane	0·41$_7$	0·44$_8$	0·42$_2$	0·50$_3$
tribromomethane	0·68$_3$	0·72$_2$	0·70$_0$	0·86$_3$
trichloromethane	0·43$_5$	0·50$_2$	0·48$_2$	0·69$_2$
ethanenitrile	0·49$_0$	0·52$_8$	0·49$_3$	0·55$_0$
propanone	0·31$_7$	0·36$_3$	0·32$_8$	0·44$_8$
dimethyl sulphoxide	0·57$_7$	0·62$_0$	0·56$_0$	0·70$_3$

TABLE 3.35

^{1}H NMR Solvent shifts of H_3CX compounds

Initial medium	Chemical shift difference/ppm	Final medium	Ref.
	1. $H_3CCHBrCH_2Br$		
cyclohexane	-0.03_3	tetrachloromethane	[263]
cyclohexane	-0.06_7	1,2-dibromopropane	[263]
cyclohexane	-0.13_3	dimethyl sulphoxide	[263]
	2. H_3CCH_2Br		
cyclohexane	0.06_7	tetrachloromethane	[263]
cyclohexane	0.00_0	bromoethane	[263]
cyclohexane	0.11_7	dimethyl sulphoxide	[263]
	3. H_3CCH_2Cl		
cyclohexane	0.11_7	tetrachloromethane	[263]
cyclohexane	0.05_0	chloroethane	[263]
cyclohexane	0.03_3	dimethyl sulphoxide	[263]
	4. H_3CCCl_3		
cyclohexane	0.11_2	tetrachloromethane	[263]
cyclohexane	0.08_3	tetrachloromethane	[391]
cyclohexane	0.09_2	1,1,1-trichloroethane	[263]
cyclohexane	0.07_8	1,1,1-trichloroethane	[391]
cyclohexane	0.21_2	dimethyl sulphoxide	[263]
cyclohexane	0.13_4	dimethyl methanamide	[391]
	5. H_3CCl		
cyclohexane	0.23_3	tetrachloromethane	[263]
cyclohexane	0.11_7	tetrachloromethane	[391]
cyclohexane	0.11_7	chloromethane	[391]
cyclohexane	0.20_4	dimethyl methanamide	[391]
cyclohexane	0.18_3	dimethyl sulphoxide	[263]
	6. H_3CBr		
cyclohexane	0.11_1	tetrachloromethane	[391]
cyclohexane	0.12_5	bromomethane	[391]
cyclohexane	0.20_6	dimethyl methanamide	[391]
	7. H_3CI		
cyclohexane	0.10_5	tetrachloromethane	[391]
cyclohexane	0.13_0	iodomethane	[391]
cyclohexane	0.16_0	dimethyl methanamide	[391]
	8. H_3CCN		
tetrachloromethane	-0.00_8	ethanenitrile	[391]
tetrachloromethane	0.12_9	dimethyl methanamide	[391]

or everybody in Shakespeare. It's this contribution of sharp *realism* to our muddled life that keeps us reading Homer, listening to Wagner, watching the plays of Euripides and Tennessee Williams.)

Since Freud at the latest, we Westerners have taken for granted the once-radical propositions that the hardest thing in the world for us to see, let alone understand, is everyday life; and that we can never hope to grasp it without guides, be they artists or psychiatrists or shamans. (Well, Heraclitus knew these curious, basic facts, but it took Freud to teach them to us all.) Whoever would find the heart's homeland, then, cannot go unaccompanied; and, very often, the guide given turns out to be a stranger—the last person, in fact, the voyager might imagine would know the way home, not being native to those parts. To find his path back to the little Southern town where he was born—into the deeper truths of it, lying behind its graceful porches and shaded streets—William Faulkner had Homer and Virgil for companions, and the European novelists of a century before; and, as all the literate world now knows, they proved true and reliable guides for Faulkner on his way home.

Whatever endures, and is right, in what I found on my own travels homeward was pointed out by strangers from times and regions of the mind lying far beyond the South. Among those I have acknowledged are Simone Weil and Martin Heidegger, the architects and poets of ancient Rome, and Susan Sontag. But there is another, whose art—made in a place very far from the Southern riversides and hills where my ancestors dwelt—helped me again and again to see what I otherwise might have missed during the seasons of my journeying. His name is Anselm Kiefer.

In 1991, almost exactly a year after Aunt Vandalia died—still bewildered by what I had discovered of myself and my family among her papers, not yet imagining a book would come from that bewilderment—I traveled to Berlin to see a large show of works by this outstanding contemporary German artist. It was a European exhibition of the sort my newspaper has often dispatched me to review during the nearly twenty years I have been working for them as an art critic. This time, however, I went on my own, prompted by an interest that had been deepening since 1987, when I first saw his work.

Born in 1945, Kiefer came to art after several false starts, including law school and monastic life. By 1968, he had decided on a career in art, which in turn plunged him into the late-sixties debate about Germany's cultural future that was raging among German artists and intellectuals. Kiefer belonged to a tormented generation—one without direct knowledge of the National Socialist era, but revolted by what they knew of the Third Reich their parents had built, sustained, fought for; and also at strong odds with the comfortably forgetful, Americanized culture of Germany during the years of *Wirtschaftswunder*. In an attempt to cleanse themselves from the smear of a malignant past, many young creators and thinkers opted for their own kind of aggressive amnesia—a forgetting of the old European high culture of art and music and philosophy Hitler's regime had terribly twisted to its own ends. Because *Die Meistersinger von Nürnberg*, for example, had been performed at Bayreuth in a hall hung with twisted crosses, and played as a oratorio to National Socialism, it had to be buried for all time. It would have been pointless to argue that nothing could be further from fascist cultural squalor than Wagner's splendidly humane and hilarious music-drama. The important fact for this generation was that Nazis had liked it. And if everything else the Nazis believed must be repudiated, their artistic taste was considered a true guide to what was hateful. Almost no artist coming of age in the 1960s was prepared to question this curious notion, and risk a peek around the lead curtain of forgetfulness that separated postwar Germany from the German cultural landscape that lay just back of it, and far beyond.

Kiefer was different—scandalously different, in the view of many critics and fellow artists then. And to this day, few critics and thinkers have been prepared to accept his profoundly conservative summons to Germany to recover remedies for healing the European wasteland from German history. Naïvely and clumsily at first, then, from the late 1970s onward, with increasing conviction and authority, Kiefer celebrated the very topics others were vociferously rejecting: the beautiful German hills and plowed fields and forests dear to the writers and painters and musicians of the nineteenth century, those old sources of an entirely justifiable love of country and language. On these depictions Kiefer has often inscribed the names of barely mentionable figures from myth and modern German history: the heroes of Teutonic epic and legend, Richard Wagner, the poet Friedrich Hölderlin, Martin Heidegger.

4 sp² Carbon

4.1 Alkenes

4.1.1 HC(X)=CYZ

Virtually the only evidence for hydrogen bonding by alkenes has been obtained from their vibrational spectra. In their survey of C–H groups as acceptors in hydrogen bonding, Allerhand and Schleyer [7] observed the C–H stretching mode in the infrared spectrum of the molecules listed in Table 4.1 (see p. 116). The wavenumber of this mode (3060–3100 cm⁻¹) was found to decrease when azine or dimethyl sulphoxide was added to a CCl_4 solution of any of these compounds, a new peak or shoulder appearing; this is evidence that a complex forms between the solute and the donor species.

The magnitudes of these frequency shifts are comparable to those of the trihalomethane molecules, as may be seen from comparison of the data in Tables 4.1 and 3.7. It is evident from the results for the first two molecules in the former Table that an electronegative substituent located *cis* to the C–H bond of interest results in a larger frequency difference on complex formation than does the same substituent located *trans*. In addition, the same substituent both *cis* and *trans* leads to a still larger frequency shift.

The substituents tested may be ordered according to the magnitudes of the C–H frequency shifts upon complex formation: $CCl_3 > Br > Cl > F$. Except for the group CCl_3, this order of substituents corresponds to the order of C–H frequency shifts in trisubstituted methanes.

4.1.2 H₂C=CYZ

The frequency of the maximum absorption of the H_2C 'wagging' mode of 10 such compounds has been measured, in a variety of solvents [99]. Although this mode is found to shift by a maximum of ca. $+27$ cm⁻¹ (Y = Z = Cl; solvent = dimethyl methanamide), separate bands from 'free' and 'complexed' molecules are not observed. The evidence would appear to be insufficient to show the presence of hydrogen bonds in such compounds.

4.2 Aromatic Compounds

Marvel, Copley and Ginsberg [260] postulated that the vinylogues of the trihalomethanes,

Anselm Kiefer has never needed reminding that the music of Wagner and much else from the high German past had been mobilized in the fascist war against Western civilization, even as love of the land had been perverted into the obscene *Blut-und-Boden* mythologies of the Nazi era. A revival of that love without a frank acknowledgment of its ruination would be not only dangerous, but false—as rankly false and futile as the fabrication of the Old South myth in the years after the Confederacy's defeat. But to pretend as though no goodness and beauty had been created in Germany's cultural experience would also be to falsify history, and truth itself. Hence his choice of subjects and allusions; hence, too, the grief that pervades Kiefer's magnificent landscapes, the infinite pathos and muted patriotism in his paintings of his homeland's open fields, poisoned and charred by vile war, but still sparkling with promises of renewal, in the light frost of morning, or the kindly wetness of recent falls of rain.

The 1991 show was a courageous tribute to a serious and immensely gifted German painter, whose work Germans did not want to see. It was also a homecoming of sorts for an artist who had earned great fame abroad, despite dislike in his native land. But quite apart from whatever the art world might make of their appearance in Berlin, the paintings I saw there penetrated deep into my own turmoil. Driving my own inward churn, I was beginning to understand, was a cycle of vexing questions about the South parallel to those prompting Kiefer to make the art he did. Given a history of grace and beauty contaminated by evil, how was I to remember it? How could I go into the poisoned past and find the good created there, without being poisoned? How was I to acknowledge virtues of land and memory not only perverted by malice and condensed into a justification for slavery—the fate of our Classical ideals in the decades before the Civil War—but also sustained after the defeat of 1865 by injustice crystallized into law and the very code of social life? Until the recent past, Southerners had by and large refused to accept *as our own doing* the cultural hallucination and extinguishing force of racism, and the poison seeping out of it into every institution and interchange throughout the history of Southern culture. How could I name and reclaim the good, without accepting the unacceptable?

Kiefer's sobering answer: I could not. No one can. The man or woman who journeys into the dry, suffering wasteland to free and reclaim the ancient waters of imagination is always at risk of pollution by the tainted ground. There is always the danger that, in finding what is worthy in that past, the seeker will be seduced by the illusions of the wasteland into some fresh, false romanticizing of what went before.

Unless—and with this, Kiefer's art became a companion in my travels through Southern history—unless the one who seeks is prepared to acknowledge, without excuse or exculpation, the utter failure of the ancient lights and wisdom to make men good, and the failure of his ancestors to walk by those lights and serve humankind according to the highest ideals and glory delivered to them. As the antebellum culture of my native land had culminated in an antiquarian racist empire, so had that of Kiefer's Romantic Germany. In his paintings of the burned fields of Brandenburg I saw the burned fields of my great-grandfather's South Carolina. In his depictions of once-splendid German temples in ruin, I saw the noble Classical architecture of the South destroyed. And though I had not smelled the acrid smoke of our Southern houses, fields, and visions in flames, I had witnessed more recent devastations brought down on the South by the unworthiness of my countrymen, of my family, of myself: the spiritual ruin inflicted by our adoption of America's materialism and will to power, the emptying of the countryside people into the huge cities of the New South, the dissolution, in my postbellum family, of old loyalties to the Southern earth and to the unfulfilled promise of our memories of Classical civilization.

It was in such an inner state of welcome disillusion that I began my trips in the South shortly after returning from Berlin. When it did occur to me that a book was to come from these experiences—the thought first came to me, as I recall, on a stormy, bitterly cold day in Toronto—I knew my writing would never be a grand prescription for the South's future. Susan Sontag, writing about the time Anselm Kiefer was making his earliest expeditions into the poisoned territory of Germany's past, tells us that "the shrewdest thinkers and artists are precocious archaeologists of these ruins-in-the-making, indignant or stoical diagnosticians of defeat, enigmatic choreographers of the complex spiritual movements useful for individual survival in an

should exhibit complex-forming properties similar to those of the corresponding trihalomethanes. In particular, the 4-hydrogen was expected to participate in hydrogen bonds with donor atoms. It was further expected that self-association would be observed when X = F (since the CF_3 group would be strongly electron donating), whereas none would be found when X = Cl.

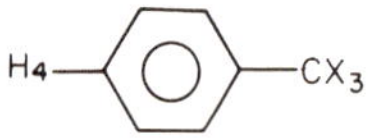

With these considerations in mind, it was felt that the enthalpy of mixing of these compounds with electron donors should be smaller for X = F than for X = Cl, since the enthalpy of self-association would offset to some extent the enthalpy of association with the donor molecule. The results in Table 4.2 (see p. 117) show that the expectation is borne out in some cases and not in others. The evidence is not conclusive regarding the nature of the interactions in these mixtures.

It has been reported [365] that the *cis* isomer of 2,3-diphenyl-1,4-dioxane exhibits a weak band in the ultra-violet spectrum at a wavelength of 315 nm; this band is not observed in the *trans* isomer, nor in 2,5-diphenyl-1,4-dioxane, nor in 1,2-diphenylethane. The band is assigned by Stumpf to a hydrogen bonded residue as shown:

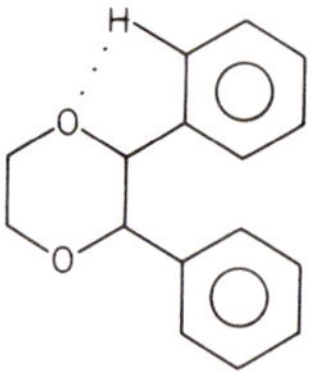

The frequency of this band is quite solvent-dependent: upon changing from 1,4-dioxane to methanol solvent, it shifts to 307 nm.

Schaefer and Schneider [335] have reported that the chemical shift difference between H_2 and H_3 in a series of 4-substituted-1-methylbenzenes is solvent dependent, being generally smaller in propanone and larger in benzene than in hexane solvent; the values are reported in Table 4.3 (see p. 117), including those for the analogous compound 4-methylazine [336]. The same authors reported similar differences in shifts of the hydrogen atoms on 5-membered unsaturated heterocyclic rings [336]. In general, the hydrogens α to the heteroatom were found to shift farther than those β to it. The resulting shift differences in the same three solvents are given in Table 4.4 (see p. 118).

A study by Dullien [111] purports to establish the presence of hydrogen bonds involving an oxole ring. The C=N stretching mode at about 1600 cm^{-1} in the Raman spectrum of compounds of the type shown appear at slightly higher frequency in the *syn* isomers than in the *anti* ones. This is attributed

syn   anti

[111] to hydrogen-bond-like interactions as illustrated. This evidence is very indirect and certainly not conclusive.

The X-ray structure of the compound (below) has been interpreted as evidence for a C–H···N hydrogen bond (H···N distance = 215 pm) [137]. However, a subsequent infrared spectroscopic study of the compound and two model compounds, in which such an interaction is expected to be absent, showed very little difference in the C–H stretching region for the three compounds [155]. Hence, the claim for such a hydrogen bond does not rest on firm ground (see Section 5.2.1).

As part of their survey of C–H groups as participants in hydrogen bonding, Allerhand and Schleyer [7] found that the C–H stretching modes of benzene and chlorobenzene are unaffected by change of medium, being found at essentially the same frequency in such diverse solvents as hexane, triethylamine, diethyl ether and azine. 1,3-Dichlorobenzene and 1,4-dichlorobenzene showed a slight modification of the C–H region in strongly donating solvents. The C–H modes of 1,3,5-trichlorobenzene and more highly chlorinated benzenes are strongly affected by the donors dimethyl sulphoxide and azine in tetrachloromethane solution, as indicated in Table 4.5 (see p. 118). The frequency shifts observed are comparable with those of trichloromethane in the same solutions.

4.3 Aldehydes

There is moderately abundant evidence that aldehydes undergo specific inter-actions with other groups (intramolecular) and other molecules (intermolecular). However, perhaps more than is the case with any other C–H-containing groups, there has been considerable controversy over the interpretation of the available evidence as to whether the aldehyde group participates as an acceptor in hydrogen bonds.

4.3.1 General Evidence of Association

Two important non-spectroscopic studies have established the phenomenon of intermolecular association involving aldehyde groups. Alexander and Lambert [4] measured the second virial coefficient of ethanal, CH_3CHO, and found that the value did not at all agree with that predicted by the Berthelot equation. They accounted for the difference in terms of association of ethanal molecules:

$$(CH_3CHO)_2 \rightleftharpoons 2\ CH_3CHO \tag{4.1}$$

$$K_\phi^{-1} = 51.6 \text{ atm at } 313.2 \text{ K}$$

The value of K_ϕ was found to be independent of pressure at this temperature. The temperature dependence of K_ϕ yields the enthalpy of complex formation: $\Delta_\phi H = -21.87 \text{ kJ mol}^{-1}$.

Postulating a cyclic form of dimeric molecule, the authors give 10.9 kJ mol^{-1} as the enthalpy of a C–H$\cdots$O hydrogen bond. This may be compared with the calculations of Morokuma [37] who finds for $H_2O\cdots HCHO$ an intermolecular binding energy of ca. 2 kJ mol^{-1}, which is expected to be smaller when H_2CO is the electron donor species.

More recently, Lee and Kumler have reported the results of a study [238] of the compound methyldiformylamine, $H_3CN(CHO)_2$. They find that, although analogous ketones have dipole moments in the ranges

cis–cis	cis–trans	trans–trans
$(8.7–9.7) \times 10^{-30}$	$(10.0–10.7) \times 10^{-30}$	20×10^{-30} C m

nevertheless, the dialdehyde has a moment of $(5.4_4–7.2_1) \times 10^{-30}$ C m in the solvents heptane, benzene and 1,4-dioxane.

era of permanent apocalypse. The time of collective visions may well be over . . . But the need for individual spiritual counsel has never seemed more acute."

If surely not among the "shrewdest thinkers" Sontag had in mind, I knew that the heyday of collective visions was indeed over, or should be: for our wretched century had known too many of them—socialist, fascist, mass-democratic, materialistic—and suffered wounds from them that will take another century to heal, if indeed the cultures of the West can ever be healed from our self-inflicted damage. And while it sounded attractive in the early 1970s, when Sontag was writing, "the need for individual spiritual counsel" had by the 1990s become, at least to me, a need worth questioning. For even as the last great Western authoritarian power, the Soviet Union, was collapsing under the weight of its own incredibility, the declining liberal-democratic collectivisms, in America and Europe, seemed to be inspiring ever-greater mass needs for authority. And, as usual, the gurus and radical rightists and television evangelists and Hollywood were wasting no time in meeting those desires with fantasies.

From the time I began this project, the industry of need-meeting was never anything I wanted to be part of. But even were I inclined to draw from my travels and thought a new "collective vision" or a package of "spiritual counsel" for the South, I could never impart any wisdom the people of my homeland might win from a renewed dialogue with Southern tradition, being now permanently a stranger to the South. More important, I could not partake in any affliction that my proposals might wreak upon the South, were they to be believed and acted upon. To attempt to write a book of advice struck me from the outset as folly, and worse: a kind of irresponsible presumption on my part, doomed to triviality, even if trivially true to the places I visited and histories I encountered. Anyway, I happened upon no bit of advice to pass on—only memories of what the Southerners in one family discovered, endured, or enjoyed, lost and rediscovered over nearly four hundred years. I went out to seek the wellsprings of my existence in the wasteland, the waters from deep time running through a Southern family, which have nourished my mind and infused the textures and inclinations of my life, for both good and ill.

But, as on every journey of the soul, what I sought at the end of the road had been there at the beginning, all along; and what I found, the stories and histories I was born with and grew up in, had never been lost. It was just that I had forgotten where to look—or, more accurately, whom to ask.

By the time I remembered, it was almost too late. The book was almost finished, my travels, I thought, done. Following my family's trace across the South and across time had taken me from the Virginia tidewater through the Carolinas to the pine-clad Texas hills around Henderson and Palestine, just east of the line where the South ends and the American West begins—the whole route along which memories had been left in the wake of my family's passage, and remained, waiting to be gathered. But along this long trail, I knew, there remained one place to go, or return to—a patch of Southern ground to which I belonged more intimately than any of the spots I had visited—but its name or location kept eluding me.

Until, that is, one blustery, very cold Toronto morning in the early spring of 1997—very like the one, some years before, on which I first knew this book would be written—when I was absentmindedly doing some household busywork, and the words *Spring Ridge* came drifting slowly up from the mental depths. At first the old phrase insisted on nothing, issued no command or summons. It was merely the property once owned by my father, then, after his death, by my sisters and me, now by someone else—or so I had long been saying to myself, whenever the name or topic came up, simply as a way to push it back into the depths for another season.

Since my late twenties, when my mind was crushed by the weight of too many fantastical pseudo-memories of the plantation South and my father, I had seldom given Spring Ridge much thought. And, indeed, during the first months and years of psychotherapy that followed that nervous breakdown, the name and place gradually sank away into the darkness—as though, in being stripped of the malignant power I'd given it, Spring Ridge had lost any right or reason to be remembered at all. On visits to Greenwood, I almost never went there. My usual excuse was that I did not want to see the house in which I spent my earliest years in grave dilapidation—a condition I blamed, at first, on the irresponsible tenants we rented it to; later, when nobody lived there anymore, on the havoc wreaked by the extreme heat and bitter cold of our upcountry

They propose association of such molecules to form dimeric complexes; since the ketones show no such complex formation, it is suggested that the formyl hydrogen atom must play a prominent part in this process. The only complex which significantly reduces the effective dipole moment is the symmetric, cyclic one involving two hydrogen bonds between the formyl groups of a pair of *cis–trans* molecules.

This proposal is consistent with the observed solvent dependence of the dipole moment, although it is not clear why the extent of dimerisation should be greater in the more strongly interacting solvent 1,4-dioxane than in the relatively inert heptane.

The NMR dilution shift of methyldiformylamine in tetrachloromethane is 0.28 ppm, comparable with that of trichloromethane [238]; this observation is consistent with a similar type of self-association interaction for the two molecules, although it is certainly no proof of this.

The molecular weight of methyldiformylamine monomer is 87, and that of the dimer 174. The molecular weight was measured at 310 K by osmometry; this gives an average value of 161 (range: 147–178). A determination at the boiling point of benzene by boiling-point elevation yields an average value of 171 (range: 153–186). Both these values indicate nearly complete dimerisation, even at the higher temperature.

If the dimer were taken to be completely symmetrical, it would have zero dipole moment and the measured values would be inconsistent with the molecular weight determinations; hence, the ring of the dimeric complex must be puckered in such a way as to result in an appreciable moment.

4.3.2 Infrared Spectroscopic Evidence

The infrared spectra of benzaldehyde and 2-hydroxybenzaldehyde have been investigated in the regions of the third and fourth overtones of the formyl C–H stretching mode [133]. These modes are found at 8793 cm^{-1} and 11 474 cm^{-1}, respectively, for benzaldehyde.

Two bands in near proximity are assigned by Freymann and Freymann [133] to hydrogen-bonded C–H groups; these appear at 7888 and 10 295 cm^{-1}. Since similar behaviour is observed for 2-hydroxybenzaldehyde, which is known not to form an intramolecular C–H···O bond, the infrared observations are taken as evidence for intermolecular association of both these compounds.

The infrared spectra, in tetrachloromethane solution, of ethanal and 4-chlorobenzaldehyde have been reported [7] in the presence and absence of dimethyl sulphoxide (0.95 mol dm^{-3}) and of azine (2 mol dm^{-3}). No change was observed in the spectrum in the region of the formyl C–H stretching band, this being taken as good evidence for the absence of hydrogen bonding by aldehydes. Similarly, the C–D stretching mode of dimethyl methanamide-d$_1$ [(CH$_3$)$_2$NCDO] is found to be unaffected by transfer from the gas phase to a 1:1 molar ratio in the solid phase with 1,3,5,7-tetranitro-1,3,5,7-tetranitra-

Southern seasons. The plain fact was that I did not wish to go because it evoked thoughts of my early, happy boyhood, and of the rupture that ripped it away from me, leaving in its wake the strange, lonely child I became and remained after my mother and I moved to Shreveport.

One of the very few times I returned to Spring Ridge was the year—I cannot now recall when—my sister Erin advised Anne and me that the state of the house had declined so irretrievably far that our only choice was to demolish it. I returned to the South not long after the demolition, and to the rubbly place on which the white frame house had stood. I did so not an hour too soon. With the same quick relentlessness that had turned the abandoned cotton fields into a nearly impenetrable thicket after the death of John Bass Mays, the green powers of the earth had already begun to swallow up the rubble, and were about to engulf the site completely.

Ranging over the scatter of detritus, I found three things to keep. One was a scrap of wallpaper my mother had hung, printed with the white blossoms and waxy leaves of *Magnolia grandiflora*, still stuck to the underside of planking. Another was a plain country brick from the foundations, that I took away and years later buried, with makeshift rites improvised from ancient Roman house-founding rituals, in a hole blasted out of the concrete footings of the old tool-and-die factory in Toronto then being converted into the dwelling place my family and I now occupy. The third was the base of a Doric column from the front porch. It is now on the deck garden I keep on my factory roof, obscured in summer by the flowerpots I set on it, abandoned to the weather in winter. The square wooden base is a thing of no beauty, certainly no value or consequence. Yet I cherish it above most things I have, only because upon it once rested a small, graceful column that, in turn, helped support a welcoming porch, in the shade of which I found refuge from the high summer heat and sat out, with my chin on my knees, the sudden downpours of Louisiana autumn afternoons.

The day I gathered up these legacies—which were precious to me then, however fraught with anxiety they might have been in earlier times—was the last time I visited Spring Ridge as an owner of it. The property, together with the house site, was sold in 1987. I learned that the pending deal was done when my dying sister Erin's husband called me in Germany, where I was reviewing an exhibition for the paper—then passed the night weeping more bitterly than I had imagined I would, or ever could. With the coming of the gray European dawn, the tears stopped, along with the surging waves of remembrance that prompted them; and in the days and weeks thereafter, I allowed Spring Ridge to slip slowly out of mind.

As the resolution to return there resurfaced, unbidden, on a snowy Toronto day ten years later, another name came to me, inextricably bound up forever with Spring Ridge. It was the name of Essie De. And a few days later, Richard Rhodes, the photographer I worked with throughout this writing, accompanied me to the front door of her house in Shreveport, the front garden resplendent with the red azaleas she loves, and tends lovingly.

Somewhere beyond her seventy-fifth birthday—which she strictly ordered me to ignore, forbidding even a card—Essie De was, of all the mammies any white Southern boy ever had, surely least like the broad-beamed, kerchief'd black women in Old South fantasy films. As she had been fifty years before, when minding me on my father's cotton plantation, so she was that day in 1997: wiry, compact, and quick, smartly dressed, and exuberantly, unsparingly plainspoken about everyone, black or white, she had known throughout her long life. As the three of us drove along the old, slow road out of the city toward Spring Ridge, De told us rapid-fire stories about the myriad living people she had always called her "black children and white children," the family of John Bass Mays, her own son and grandchildren and their numerous relations, and some very funny tales about those long dead.

Nearing Spring Ridge, however, the pace of her storytelling slowed; and when we reached the crossroads that had once defined a corner of my father's land—now being sold off, lot by lot, for the construction of bungalows—De began to reminisce about people and places I barely remembered, if at all. There was Willis Hill, the white overseer of Spring Ridge plantation, heroically handsome on horseback in John Mays's hunting photographs from the 1930s. There was Bush, the mighty black man who shouldered the practical business of planting and harvesting the cotton. Stepping forward from her stories, too, were the shadowy figures of Miss Lurleen and her husband, Mr. Sam Hall, our neighbors, whose only child died in the Sec-

cycloöctane [40]. Again, this behaviour suggests the absence of aldehydic hydrogen bonding in this system.

However, the formyl stretching band of trichloroethanal, CCl_3CHO, which is expected to have a more acidic hydrogen atom, is found at 2938 cm^{-1} in CCl_4 solution, but at 2805 cm^{-1} in a CCl_4 solution of azine [379]. This behaviour may be indicative of hydrogen bond formation. The temperature dependence of the frequency shift gives an enthalpy of formation of (10 ± 2) kJ mol^{-1}.

Considerable controversy has arisen over the possibility of hydrogen bond formation in 2-substituted benzaldehydes. Pinchas has postulated, and later argued at some length, that the evidence from the infrared spectra of these compounds proves unambiguously that such hydrogen bonding occurs [310, 311, 312, 313]. The principal basis of this contention is that in *all* 3- and 4-

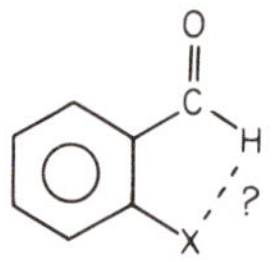

substituted benzaldehydes, the formyl C–H stretching mode falls within the narrow range of 2720–2745 cm^{-1}. On the other hand, a donor substituent (Cl, Br, NO$_2$, OCH$_3$) in the 2-position is almost invariably accompanied by a C–H stretch at 2747–2765 cm^{-1}, i.e. at higher wavenumber. The fact that 2-methylbenzaldehyde displays a 'normal' C–H stretch at 2725 cm^{-1} is regarded as confirmation that steric crowding is not an important factor. The unperturbed behaviour of ν_{CH} in 2-hydroxybenzaldehyde is, then, consistent with the established structure shown, which admits of no C–H···O bond. The non-hydrogen-bonded behaviour of benzene-1, 2-dicarboxaldehyde is not easy to reconcile with the hydrogen bonding alleged in the 2-nitro compound.

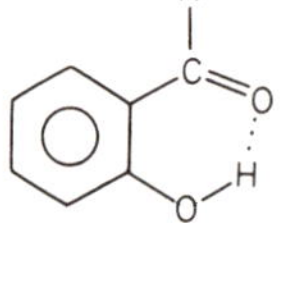

West and Whatley [395] have pointed out a couple of weaknesses in the Pinchas model. The first is that the N–O stretching mode of a nitro group is usually observed at lower wavenumber when participating in a hydrogen bond. However, it is seen that the 2-nitro compound shows no decrease in ν_{NO} relative to the other isomers.

	2-NO$_2$	3-NO$_2$	4-NO$_2$
ν_{NO}/cm^{-1}	1340	1345	1336

Second, ν_{CH} in the molecule 2,4,6-trimethylbenzaldehyde is found at 2761 cm^{-1}, which is well within the range which is said to indicate hydrogen bonding. Pinchas attempted to explain this observation as due to the presence of quinoid structures such as that shown. However, Forbes pointed out [129] that analogous structures would be expected for other 4-substituted benzaldehydes such as 4-nitro, but are not observed.

Forbes [129] observes the symmetrical N–O stretch in a nitrobenzene to be unaffected by the presence of a 2-formyl substituent, but displaced to lower wavenumber by a 2-hydroxy group. He concludes that the formyl group does not participate in intramolecular hydrogen bonding with the nitro group, and that the behaviour of the C–H bands is consistent with steric interactions. He mentions,moreover, that the difference in CHO chemical shift between 2-substituted and 3- or 4-substituted benzaldehydes (0.4 ppm) is considerably too small to be due to hydrogen bonding, which usually leads to ^{1}H shifts of several ppm. In addition, it is found [219] that this chemical shift difference is nearly constant irrespective of the nature of the substituent.

Schneider and Bernstein [339] have compared the gas- and solid-phase infrared spectra of methanal, HCHO. The C–H stretch is found at *higher* wavenumber, and the CH$_2$ deformation mode at *lower* wavenumber in the solid phase spectrum. They take these changes, together with a reduction in $\nu_{C=O}$ as evidence of strong intermolecular interactions; however, since the CH$_2$ frequency shifts are in the directions opposite from those characteristic of hydrogen bonding, they conclude that this is not the mechanism of interaction. It may be borne in mind that the frequency changes observed by Pinchas in 2-substituted benzaldehydes are also to higher wavenumber, which is behaviour not normally considered to be characteristic of hydrogen bonds.

It appears that the process of solidification of methanal differs from association to form dimers, since it is found [169, 215] from matrix isolation studies that the C–H stretching and CH$_2$ deformation modes *both* shift to a higher frequency in the dimer. Khoshkhoo and Nixon suggest that the dimer has a centre of symmetry involving the carbonyl groups, and assert [215] that the observed frequencies are 'incompatible with a hydrogen-bonded model of the dimer'.

Lesch and Ulbrich [240] find that the C–H stretching mode of methanoic acid shifts to *lower* wavenumber when that substance is dissolved in dimethyl sulphoxide ($\Delta\nu_s = -39$ cm^{-1}) or in azine ($\Delta\nu_s = -57$ cm^{-1}). These shifts in ν_s are coincidentally almost identical to those of trichloromethane in the same solvents (-41 and -54 cm^{-1}, respectively). However, it will be recalled that the spectrum of the similar molecule ethanal undergoes no change in concentrated CCl$_4$ solutions of the same donor substances. Evidently, we must postulate that, either 'activation' of C–H by the OH substituent is a prerequisite for association interactions, or, dilution in tetrachloromethane breaks up all the association. In the latter case, which seems the more likely, the interaction must be described as a general solvation rather than a specific association. On the other hand, perturbations of ν_{CH} in methanoic acid may also be due to perturbations of the OH group; the rubidium salt exhibits this mode displaced by -143 cm from the acid [240]; hence, it is not necessary to postulate C–H hydrogen bonding in this case.

Wilmshurst has observed [301] that the out-of-plane C–H bending mode of methyl methanoate, HCOOCH$_3$, in CS$_2$ or CCl$_4$ solution, exhibits two bands, one of which is attributed to a complex on account of its concentration dependence. The nature of the interaction, is, however, unknown.

4.3.3 NMR Spectral Evidence

The formyl chemical shifts of a number of aldehydes have been measured for the pure liquids [328]. None of these shifts give any indication of intermolecular association, apparently being accounted for entirely in terms of intramolecular electronic effects.

TABLE 4.1

Infrared spectral frequency shifts of substituted alkenes upon formation of complexes in CCl$_4$ solution, relative to the C–H stretching frequency [7]

Compound	$10^4 \, \Delta\nu_{CH}/\nu_{CH}$	
	0.95 mol dm^{-3} dimethyl sulphoxide	2 mol dm^{-3} azine
HC(Cl)=CHCl [*trans*]	107	110
HC(Cl)=CHCl [*cis*]	58	75
HC(F)=CF$_2$		71
HC(Cl)=CCl$_2$	133	146
HC(Br)=CBr$_2$	130	170
HC(CCl$_3$)=CCl$_2$		208

ond World War, leaving them to live out their lonely days behind drawn shutters in the house just down the road from our own; and less tragic folk, black and white, whose lives had been woven together by time's shuttle on the cotton farms that once fell away in four directions from the crossroads at Spring Ridge. All these people were gone, along with the farms, now buried deep by heaps of briar and stands of scrubby trees and impenetrable underbrush, at least beyond the small, intermittent plots cleared for house building. No trace remained, of course, of our own home, or the store my father kept, or the barn where John Mays stabled his mules—but Essie De remembered where they had stood, and pointed out the sites, rekindling in me, with each gesture, thoughts of the boyhood I had lost, now untroubled after the smoothing passage of so many years.

We walked the margin of my childhood homeplace, to the bridge under which Spring Branch still flowed toward the unfathomable Blue Hole, and the depths into which De, her son, and I had long ago dropped our hooks, then waited in the afternoon stillness and heat for the tug on the end of the line. And a little later, we walked just beyond the western edge of the land, to the cemetery on the ridge from which our plantation took its name—the long rise in the pattern of undulating hills lying toward Texas, dotted by small springs feeding branches that run eastward in gullies ever deepening and broadening as they cross what once were cotton fields, disappearing at last into the sprawling swamps and deep bottomlands along the Red River.

Under the hillside cemetery's green furze and shade, in graves marked and unmarked, lie Essie De's people, her mother and kin and ancestors—once sewn into a black fabric that paralleled the white one woven by my family, and, all of us, black and white, stitched into the ample, embracing quilt of land. De and I strolled the black cemetery's grounds; and, as we did, she told still more stories— of how, as a girl, she often came up to the cemetery to think and read; how, when pregnant with her only child, she came there just to be alone with the trees and flowers—each story a thread in the human skein I imagined, when a child, to be eternal.

But then my mother and I went away, I told her; and the fabric was broken.

We all went away, she replied—then added, with a bright

TABLE 4.2

Enthalpies of mixing of compounds $C_6H_5CX_3$ with donor molecules at 276 K in equimolar proportions [260]

Donor	$-\Delta_m H/\text{J mol}^{-1}$	
	X = F	X = Cl
dimethyl ethanamide	525	1275
cyclohexyldimethylamine	500	1715
triethoxyphosphine oxide	565	670
diethyl ether	585	515
propanone	165	165

TABLE 4.3

^{1}H NMR Chemical shift differences of H_2 and H_3 in 5 mol % solutions of aromatic compounds [335]

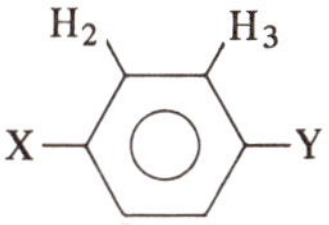

Substituent		$(\delta_3 - \delta_2)/\text{ppm}$		
X	Y	propanone	hexane	benzene
NO_2	CH_3	0·598	0·818	1·203
SO_2Cl	CH_3	0·403	0·517	1·012
CH_3CO	CH_3	0·577	0·650	0·833
NC−	CH_3	0·260	0·225	0·272
F	CH_3	−0·192	−0·178	0·000
Cl	CH_3	0·088	0·157	0·385
Br	CH_3	0·270	0·365	0·590
I	CH_3	0·597	0·677	0·917
NO_2	F	0·503	0·512	0·473
SO_2Cl	F	0·670	0·863	1·133
Br	F	0·457	0·520	0·543
(4-methylazine)[336]		1·2$_7$	1·4$_7$	1·8$_7$

chuckle, that she had been watching me go away all my life. Not understanding what she meant, I asked her to tell on. What Essie De then said, calling up things I had not thought of in fifty years, went like this.

From the earliest time in my life until my father's death, when I was six, she and I took trips. While sitting by the Blue Hole or on the front porch, I would name a place I'd heard of somewhere, on the radio perhaps, or from the menfolk who came to the house in the evening to smoke and talk cotton. New York was one such place-name she recalled. Africa, Paris, and Berlin were others—though we could not go to Berlin, because a war was on. But if nothing so grim as war prevented our departure, Essie De would take the name, then, spinning a tale around it, take me there.

On one of our several voyages to Africa, we went out on a big safari to hunt lions and tigers, and, on another, encountered an entirely horrible tribe of cannibals who could not decide which of us, their white captive or their black, would be tastier—so, while they were arguing, we made good our escape. Paris was a large city in which stylish people lived, and a place only the most stylish people from Spring Ridge visited. So, when De and I went to Paris, we were very stylish indeed, staying in a deluxe hotel full of prompt, attentive servants (whose lot was to be bossed around by Essie De), shopping in all the best department stores for my fine clothes and De's gold jewelry until I was tuckered out and ready to go back to Spring Ridge for a nap.

New York was a favorite destination, so De took me there often—a magical and huge city, full of buildings taller than the tallest trees I had ever seen, and with more people than I had ever encountered, even on shopping days in Shreveport. There were many elevators in New York, which we rode up and down whenever we liked—so many, in fact, and so high, that I would often nod off to sleep when on the long zoom up or on the long zoom down. And when I did, Essie De would gather me up in her arms and tote me off to bed in my father's house, where I dreamed of elevators and tigers until waking, ready for her to take me on the next journey from Spring Ridge into the world beyond the cotton fields and the distant, low horizon.

As Essie De and I paced off the cemetery rows, with Richard taking pictures a ways off, at the boundary where it met the brushy for-

Expressway, Shreveport

TABLE 4.4

Chemical shift differences in some heterocyclic compounds as a function of solvent [336]

Compound	$(\delta_1 - \delta_2)$/ppm		
	propanone	hexane	benzene
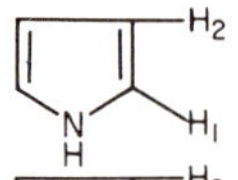	0·685	0·372	0·052
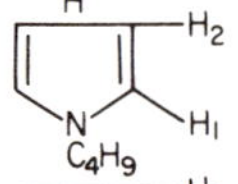	0·677	0·455	0·143
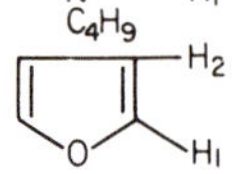	1·128	1·042	0·993
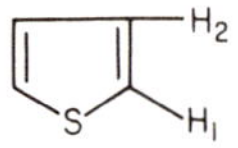	0·315	0·187	0·145
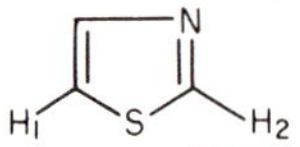	1·388	1·502	1·662
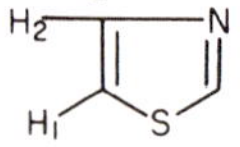	0·313	0·670	1·060

TABLE 4.5

The effect of complex formation upon the C–H stretching mode of aromatic compounds [7]

Compound	$10^4 \, \Delta\nu_{CH}/\nu_{CH}$	
	DMSO[a]	azine[b]
1,3,5-trichlorobenzene		110
1,2,4,5-tetrachlorobenzene	129	135
1,2,4,5-tetrabromobenzene		165
1-nitro-2,3,5,6-tetrachlorobenzene		146
3,5-dinitro-2,4,6-trichlorobenzene	152	171

[a] 0·95 mol dm^{-3} dimethyl sulphoxide in CCl$_4$.
[b] 2 mol dm^{-3} azine in CCl$_4$.

est, I wondered whether I had ever really been as rooted in this Louisiana ground as I later came obsessively to imagine—whether I could have stayed there, even had my father lived on past 1947 and reared me to follow his Southern path, as cotton planter and dweller in the deep countryside. For, as De brought more strongly to mind with each new reminder of our long-past imaginary trips, I had always been leaving Spring Ridge and the South. It was only a matter of time, surely, before the travels would become real, taking me at last outside the South in fact, and, in the end, forever, as they did.

On the short road back to Shreveport, and on the farther stretch of highway to Dallas and our flight home, I realized De's stories at Spring Ridge had disclosed to me a truth that had eluded me since my returns to the South began, not long after Aunt Vandalia's bequest of papers and memories had stirred in me the desire to find the South, on the land and in myself. This truth—the message given me when I finally returned to Spring Ridge and the beginning, with the surest, beloved guide to the beginning of the child I was and the man I would become—was that the summons of the ancestral voices was not to refashion bonds to the Southern land, but to acknowledge that I never possessed them; and to say the last of the innumerable farewells I was making since my first trip with Essie De to the far realms of the imagination shaped at Spring Ridge.

By 1990, the time had come for me to "revalue our secret hoard in the light of both past and future," as Hilda Doolittle wrote—to go back to the South as instructed scribe, seeking what treasures, new and old, lay in the land of my ancestors. This I did, in homage to the blood, truth, and existence I have received from my family, recovering wealth far greater than I expected to retrieve. I found precious traces of noble, failed attempts to raise on Southern ground a culture rooted in the natural orders of our seasons, to build a civilization free of cruel utopianism and metropolitan alienation, sustained by loyalties to place, and to whatever is virtuous and true in the traditions we have inherited from the West. And in the end, I found the Southerner I was and always will be: destined to live as a sojourner among nations far from my native land, yet sheltering in my heart the South's ancient lights until day ends, and my walking of all paths is done.

5 sp Carbon

5.1 Methanenitrile, HCN

It is well known that molecules of methanenitrile undergo considerable self-association, particularly in the condensed phases but also to some extent in the vapour phase. The molecular dipole moment is rather large, 9.780×10^{-30} C m [360], and is considered to be indicative of a considerable contribution to the electronic structure of the resonance form II. Indeed, the H–C bond is also

$$H-C\equiv N \qquad H-\overset{+}{C}=\overset{-}{N}$$
$$\text{I} \qquad\qquad \text{II}$$

polarised to a considerable extent, resulting in the large H–C bond moment of 3.80×10^{-30} C m [187, 371] (hydrogen positive). Hence, end-to-end interaction of HCN molecules is to be expected. The resultant C≡N bond length, 115·3 pm [319], is nearly identical to that for other nitrile bonds [175].

5.1.1 Crystal Structures

Methanenitrile is found by X-ray crystallography to exist in infinite chains of molecules in an end-to-end configuration [112], in both the high-temperature and the low-temperature modification, which differ only slightly in their packing arrangements. The C–H···N distance is found to be 318 pm [112]; if the C–H distance is assumed to be the same as in the gas phase, 106·6 pm [319], then the H···N distance would be about 211 pm, which is extremely small. However, the C–H distance is likely to increase upon hydrogen bond formation, which would make the H···N distance even smaller. When compared with the hydrogen bond distances in complexes of other molecules [60, 122, 123] (e.g. see Section 5.2.1), the C–H···N bond in solid methanenitrile appears to be quite strong.

5.1.2 Permittivity

The permittivity of HCN over a wide range of temperatures, in both the solid [361] and the liquid [80] phase, has been measured. The solid phase permittivity, ϵ_r, falls in the range 2 to 3 and is evidence for the lack of molecular rotation in the crystal (both modifications). Liquid-phase HCN, by contrast, exhibits a very

E·

high permittivity which is highly temperature dependent (see Table 5.1, p. 131). Coates and Coates [80] postulated that the high value and its temperature sensitivity could be accounted for by the formation of linear chains of HCN molecules, the average length of such chains decreasing at higher temperatures.

5.1.3 Chain Length in the Liquid

Pauling has pointed out [297, p. 458] that the observed permittivity of HCN at ordinary temperatures (ca. 295 K) is about three times as great as expected for a non-associated liquid of the same molecular dipole moment. He draws the conclusion that the HCN molecules are present in linear chains $(HCN)_n$, where the average value of n is about 3.

Tyuzyo has obtained estimates of the chain length from the viscosity [386] of liquid HCN and its energy of vaporisation [387]; these are, respectively, 3·8 and 3·1, and provide confirmation of Pauling's estimate.

5.1.4 Complex Formation Constants

The formation constants of several gas-phase complexes of HCN have been measured; these are given in Table 5.2 (see p. 132), including those for dimer and trimer formation. It is evident that the extent of complex formation in the vapour phase is markedly dependent on the nature of the donor, increasing in the order

nitrile $<$ ether $<$ amine.

No values have been reported of HCN complex formation constants in the liquid or dissolved state, although the product KC_0 (where C_0 is the stoichiometric concentration of HCN) has been calculated from the permittivities of liquid HCN [81]; this calculation was carried out making use of the assumption that K has the same value for each step in the multiple association of methanenitrile molecules. In fact, however, it is unlikely that all the K's are equal, for the

T/K	259·9	268·2	278·2	288·2	298·9
KC_0	15·6	11·4	8·26	6·20	4·65

following reasons: first, the vapour-phase values of K for the formation of dimer and of trimer are not equal, the latter being considerably the larger (see Table 5.2); indeed, it has been argued [333] that K for dimer formation *must* be smaller than that for formation of higher complexes since the former step involves a loss of entropy due to two monomers' inability to re-orient freely, whereas the latter results in loss of such entropy by one monomer only. Second, as is pointed out in the next section, the enthalpy of formation of trimer is more than twice as great as that of dimer, although the formation of twice as many hydrogen bonds is being considered.

Sources

I. GENERAL RESOURCES

Matthew Page Andrews. *Virginia: The Old Dominion* (Garden City: Doubleday, Doran & Co., 1937).

Richard R. Beeman. *The Evolution of the Southern Backcountry: A Case Study of Lunenburg County, Virginia 1746–1832* (Philadelphia: University of Pennsylvania Press, 1984).

Warren M. Billings, ed. *The Old Dominion in the Seventeenth Century: A Documentary History of Virginia, 1606–1689* (Chapel Hill: The University of North Carolina Press, 1975).

Bradley G. Bond. *Political Culture in the Nineteenth-Century South: Mississippi, 1830–1900* (Baton Rouge & London: Louisiana State University Press, 1995).

Carl Bridenbaugh. *Cities in the Wilderness: The First Century of Urban Life in America* (London: Oxford University Press, 1971).

———. *Jamestown 1644–1649* (London: Oxford University Press, 1980).

T. H. Brien. *Tobacco Culture: The Mentality of the Great Tidewater Planters on the Eve of Revolution* (Princeton: Princeton University Press, 1985).

Philip Alexander Bruce. *Economic History of Virginia in the Seventeenth Century* (New York: MacMillan, 1896).

———. *Institutional History of Virginia in the Seventeenth Century* (New York and London : G. P. Putnam's Sons, 1910).

Martin J. Dain. *Faulkner's County: Yoknapatawpha* (New York: Random House, 1964).

Donald Dudley. *Roman Society* (London: Penguin Books, 1970).

Charles J. Farmer. *In the Absence of Towns: Settlement and Country Trade in Southside Virginia, 1730–1800* (Lanham, Md.: Rowman & Littlefield Publishers, 1993).

Federal Writers' Project of the Works Progress Administration. *Mississippi: The WPA Guide to the Magnolia State* (Jackson & London: University Press of Mississippi, 1988).

. *Virginia: A Guide to the Old Dominion* (Richmond: Virginia State Library and Archives, 1992).

5.1.5 Thermodynamic Parameters

The enthalpy and entropy of formation of HCN complexes fall in to the same range as those of other C–H acceptor molecules (see Sections 3.1.4, 3.2.5, 3.4.1.5, 3.5.2, 4.3.1, 5.2.4). The vapour-phase pressure measurements of Giauque and Ruehrwein [143] yield the following values:

	$2 \text{ HCN} \rightleftharpoons (\text{HCN})_2$	$(\text{HCN})_2 + \text{HCN} \rightleftharpoons (\text{HCN})_3$
$\Delta_\phi H / \text{kJ mol}^{-1}$	$-13 \cdot 7$	$-22 \cdot 8$
$\Delta_\phi S / \text{J K}^{-1} \text{ mol}^{-1}$	$-69 \cdot 0$	$-90 \cdot 0$

Jones, Seel and Sheppard [194] obtained a value of the vapour-phase enthalpy of formation of $(\text{HCN})_2$ which is $(-23 \cdot 8 \pm 2 \cdot 1) \text{ kJ mol}^{-1}$ using infrared spectroscopy. This value is in good agreement with that from pressure measurements for trimer formation, but not with that for formation of dimer.

Similar inconsistency is found for the enthalpy of association of methanenitrile in the liquid phase. Cole, using his assumption of K's all equal, finds $\Delta_\phi H = -19 \cdot 2 \text{ kJ mol}^{-1}$ [81], whereas the measurements of Tyuzyo [387] yield $\Delta_\phi H = -29 \cdot 6 \text{ kJ mol}^{-1}$.

Several values of enthalpy of phase transition of HCN are listed here:

$$\Delta_s^l H = -8 \cdot 41 \text{ kJ mol}^{-1} \text{ [143]}$$

$$\Delta_s^v H = -35 \cdot 1 \text{ kJ mol}^{-1} \text{ [185]}$$

$$\Delta_l^v H = -25 \cdot 2 \text{ kJ mol}^{-1} \text{ [143]}$$

$$\Delta_l^v H = -28 \cdot 0 \text{ kJ mol}^{-1} \text{ [185].}$$

5.1.6 The Vibrational Spectrum

All three vibrational modes of methanenitrile are affected markedly upon formation of a complex, whether by self-association or otherwise. Table 5.3 (see p. 132) contains a listing of the observed wavenumbers of HCN. The values observed in the vapour phase correspond very closely to those for the monomeric molecule, since the extent of association in this phase is small [143]. A pronounced deuterium isotope effect is observed, even with ν_1, which involves primarily the stretching of the C≡N bond.

The vibrational spectrum of crystalline HCN and DCN has been measured, both of the low-temperature modification (I) ($\gtrsim 170$ K), and of the high-temperature one (II). The wavenumbers of the bands in the two forms are similar, with the bending mode appearing at slightly higher wavenumber in (II) than in (I), and ν_3 at lower wavenumber. The C–H stretching mode, ν_3, is displaced ca. -180 cm^{-1} on passing from the monomeric form to the polymeric. Further, the HCN bending mode, ν_2, is shifted by ca. $+115 \text{ cm}^{-1}$, while the

Richard M. Gummere. *The American Colonial Mind and the Classical Tradition: Essays in Comparative Culture* (Cambridge, Mass.: Harvard University Press, 1963).

Charles E. Hatch, Jr. *The First Seventeen Years: Virginia, 1607-1624* (Charlottesville: University Press of Virginia, 1957).

———. *Jamestown, Virginia: The Townsite and Its Story* (Washington: United States Government Printing Office, 1955).

Charles Hudson & Carmen Chaves Tesser, eds. *The Forgotten Centuries: Indians and Europeans in the American South, 1521–1704* (Athens & London: The University of Georgia Press, 1994).

Anne Lash Jester. *Domestic Life in Virginia in the Seventeenth Century* (Williamsburg: Virginia 350th Anniversary Celebration Corp., 1957).

Annie Lash Jester and Martha Woodroof Hiden. *Adventurers of Purse and Person: Virginia, 1607-1624/5* (Richmond: Dietz Press, 1987).

Frederick R. Karl. *William Faulkner: American Writer* (New York: Weidenfeld & Nicolson, 1989).

E. Kay Kirkham, ed. *A Genealogical and Historical Atlas of the United States of America* (Logan, Utah: Everton Publishers, 1976).

Katharine Du Pre Lumpkin. *The Making of a Southerner* (Athens and London: The University of Georgia Press, 1991).

David L. Madsen. *Early National Education, 1776–1830* (New York: John Wiley & Sons, 1974).

George Carrington Mason. *Colonial Churches of Tidewater Virginia* (Richmond: Whittet and Shepperson, 1945).

Jack McLaughlin. *Jefferson and Monticello: The Biography of a Builder* (New York: Henry Holt and Company, 1988).

Richard A. McLemore, ed. *A History of Mississippi* (Hattiesburg: University and College Press of Mississippi, 1973).

William Meade. *Old Churches, Ministers and Families of Virginia* (Philadelphia: J. B. Lippincott, 1861).

Robert Mills. *Mill's Atlas of the State of South Carolina, 1825* (Easley, SC: Southern Historical Press, 1980).

Christopher Morris. *Becoming Southern: The Evolution of a Way of Life, Warren County and Vicksburg, Mississippi, 1770-1860* (New York & Oxford: Oxford University Press, 1995).

Frederick Law Olmsted. *The Cotton Kingdom: A Traveller's Observerations on Cotton and Slavery in the American Slave States* (New York: Alfred A. Knopf, 1953).

John Solomon Otto. *The Southern Frontiers, 1607-1860: The Agricultural Evolution of the Colonial and Antebellum South* (New York: Greenwood Press, 1989).

Charles O. Paullin, ed., John K. Wright. *Atlas of the Historical Geography of the United States* (Washington: Carnegie Institution of Washington and the American Geographical Society of New York, 1932).

Peggy W. Prenshaw and Jesse O. McKee, eds. *Sense of Place: Mississippi* (Jackson: University Press of Mississippi, 1979).

George C. Rable. *But There Was No Peace: The Role of Violence in the Politics of Reconstruction* (Athens: University of Georgia Press, 1984).

Meyer Reinhold. *Classica Americana: The Greek and Roman Heritage in the United States* (Detroit: Wayne State University Press, 1984).

James L. Roark. *Masters Without Slaves: Southern Planters in the Civil War and Reconstruction* (New York: W. W. Norton & Co., 1977).

Conway Robinson & R. A. Brock, eds. *Abstract of the Proceedings of the Virginia Company of London 1619–1624* (Richmond: Virginia Historical Society, 1889).

Robert Blair St. George, ed. *Material Life in America, 1600–1860* (Boston: Northeastern University Press, 1988).

Charles Reagan Wilson & William Ferris, eds. *Encylopedia of Southern Culture* (Chapel Hill & London: The University of North Carolina Press, 1989).

Jack Case Wilson. *Faulkners, Fortunes and Flames* (Nashville: Annandale Press, 1984).

II. NOTES

Introduction

p. 1   *"The oldest of the old . . . ":* my translation of Martin Heidegger, *Aus der Erfahrung des Denkens* (Pfullingen: Verlag Günther Neske, 1986), 19.

p. 24   *"now is the time to re-value / our secret hoard . . . ":* Hilda Doolittle, *Trilogy* (New York: New Directions Books, 1973), "The Walls Do Not Fall," poem 36.

Chapter One

p. 30   *"The study of history . . . ":* Livy, *The Early History of Rome*, tr. Aubrey de Sélincourt (London: Penguin Books, 1960), 34.

p. 37   *Classical architecture, wrote Nietzsche:* in *Human, All-Too-Human,* quoted by George Hersey, *The Lost Meaning of Classical Architecture* (Cambridge and London: The MIT Press, 1988), epigraph.

p. 37   *Xenephon:* quoted in Bertram Wyatt-Brown, *Southern Honor: Ethics and Behavior in the Old South* (New York & Oxford: Oxford University Press, 1982), 36.

p. 42ff   *"Judge Edward Mayes, Esquire":* biographical facts from Steven Griffin Mays, "First Researchers of the Family," *Mays Research Letter,* Summer, 1994, 1–2.

p. 45   *The earliest evidence: L. Q. C. Lamar, His Life, Times, and Speeches 1825–1893.* (Nashville: Publishing Company of the Methodist Church, 1896).

p. 46   *a recent scholarly biography:* James B. Murphy, *L. Q. C. Lamar: Pragmatic Patriot* (Baton Rouge: Lousiana State University Press, 1973), 273.

p. 48ff   *Edward's* Partial History*:* Edward Mayes, *A Partial History of the Mayes Family of Virginia and Kentucky* (Jackson: Hederman Brothers, n.d. [c.1906, reprinted 1926]).

C≡N stretch is practically unaffected. This behaviour is generally considered to
be characteristic of hydrogen bond formation. The doubling of the bending
mode in the solid phase has been attributed to (a) the removal of its degeneracy
in the crystal [176], and (b) the fact that the HCN crystal is piezoelectric [305].

In the liquid phase, the observed wavenumbers are intermediate between those
of the vapour- and solid-phase molecules, although the HCN is known to be
present predominantly in the form of oligomers $(HCN)_n$, where n is a small
integer [81]. Hence, the further displacement of these modes in the solid
phase is indicative of a further influence on them by HCN molecules more remote
in a chain, and/or the influence of neighbouring chains of molecules; the latter
suggestion seems the less likely.

King and Nixon [216] have measured the vibrational spectra of $(HCN)_n$,
n = 1, 2, 3, . . ., when isolated in matrices of A, N_2 and CO at 20·5 K. The
bands assigned by them to monomeric HCN have wavenumbers similar to those
of vapour-phase molecules, although the trends in v_2 and v_3 suggest that a small
amount of interaction with the matrix may be present, particularly in N_2 and
CO. The bands which they assign to $(HCN)_2$, and which may also be due to the
trimer, are found at wavenumbers quite close to those observed in liquid-phase
HCN; this assignment, then, is consistent with the postulate of short-chain
oligomers in the liquid phase.

The vibrational spectrum of methanenitrile in complexes with other molecules
is also known, and is given in Table 5.4 (see p. 133). As is the case with self-association, the formation of complex (in the vapour phase) results in a decrease of the
C–H stretching frequency; in the solid phase, there is a larger decrease in $v(CH)$,
larger even than for the self-association of HCN. For the dimethyl ether complex,
it is postulated [344] that the complex stoichiometry is 1:1, as is also the case
for the methyleneimine complex.

Caldow and Thompson [64] found that the mode v_3 and, to a smaller degree,
v_1, decrease in frequency upon transfer from the vapour phase into solution. The
magnitude of the decrease depends upon the nature of the donor moiety in
the solvent molecule, as follows: Cl < Br < I < aromatic ring < NO_2 < C≡N <
C=O, and is not a simple function of the polarity of the solvent molecules (see
Table 5.5).

In addition, the bending frequency increases with solvent donor ability:

Solvent:	vapour	CS_2	$HCBr_3$	H_2CBr_2	$C_2H_5NO_2$	CH_3NO_2
v_2/cm^{-1} [64]	712	725	733·5	738	756	760
Calc. [2]		724	728	732	736	737

These sets of observations are consistent with hydrogen bonding between
methanenitrile and solvent molecules. Akopyan and Bakhshiev calculated [2]
the value of v_2 in these solvents on the premise that van der Waals forces alone
are responsible for solute–solvent interactions of HCN; as may be seen, their

p. 48   *Mircea Eliade:* in *Myths, Dreams and Realities: The Encounter between Contemporary Faiths and Archaic Realities,* tr. Philip Mairet (New York and Evanston: Harper & Row, 1960), 31.

p. 50   *"Nature loves to hide":* tr. Guy Davenport, *7 Greeks* (New York: New Directions Books, 1995), 159.

p. 51   *"We do not notice . . . ":* tr., Davenport, *7 Greeks,* 170.

C h a p t e r   T w o

p. 63   *another account:* Jewell Mayes, "The Mayes-Mays Family Lines," ed. Steven Griffin Mays, *Mays Research Letter* (August, 1993), 4.

p. 77ff   *In the long narrative of our relation with Nature:* The crucial book in all my thinking about this narrative is Robert Pogue Harrison, *Forests: The Shadow of Civilization* (Chicago and London: University of Chicago Press, 1992).

C h a p t e r   T h r e e

p. 97ff   *Learning was not new in the Virginia Colony:* The definitive work on seventeeth-century tidewater knowledge is Richard Beale Davis, *Intellectual Life in the Colonial South 1585-1763* (Knoxville: University of Tennessee Press, 1978). See especially Vol. II, 506, 507–8; 594–5.

p. 101   *"The colloquial word-of-mouth . . . ":* Eric A Havelock, *Preface to Plato* (Oxford: Basil Blackwell, 1963), 106.

p. 101ff   *In a brilliant aside:* Claude Lévi-Strauss, tr. John Russell, *Tristes Tropiques* (New York: Atheneum, 1963), 286–310.

p. 102   *"I am inclined to think he was right . . . ":* Lévi-Strauss, *Tristes Tropiques,* 390–391.

p. 103ff   *Hicks's* Peaceable Kingdom: see Eleanore Price Mather and Dorothy Canning Miller, *Edward Hicks: His Peaceable Kingdom and Other Paintings* (Toronto: Cornwall Press, 1983).

C h a p t e r   F o u r

p. 112   *the enterprising figure of William Gooch:* see Marian Asher Fawcett, "Historical Sketch," in The Historical Committee of the Bicentennial Commission of Campbell County, Virginia, ed., *Lest It Be Forgotten: A Scrapbook of Campbell County Virginia* (n.p.: Altavista Printing Co., 1976), 9–14.

p. 114   *the name* Staunton: Herman Ginther, *Captain Staunton's River* (Richmond: The Dietz Press, 1968), 1–2.

p. 115   *". . . neither grass could grow . . . ":* Fawcett, *Lest It Be Forgotten,* 11.

pp.115–116   *William Irvin, descendant of an early planter:* quoted in Ginther, *Captain Staunton's River,* 3–4.

p. 123ff   *The most famous of them all was Patrick Henry:* biographical information from Henry Mayer, *A Son of Thunder: Patrick Henry and the American Republic* (New York & Toronto: Franklin Watts, 1986).

p. 124   *. . . the river valley was "wild and desolate" . . . :* George Morgan, *Patrick Henry,* quoted in Ginther, *Captain Staunton's River,* 38, 39.

calculated values become progressively worse as the solvent donor ability increases.

The intensities of the bands of HCN and DCN in a variety of solvents have been reported [63, 144]; these are given in Table 5.6 as ratios with the vapour-phase values of the respective bands. Caldow *et al.* [63] point out that the intensities do not show a correlation with solvent permittivity; hence, at least part of the intensity enhancement must be due to specific interactions between solute and solvent molecules, presumably via hydrogen bonding. The intensity changes in the aromatic solvents show that the interaction of HCN is with their π electrons.

Girin, Akopyan and Bakhshiev [144] have calculated the intensities of these bands to be expected on the basis of the assumption that HCN interacts only via van der Waals forces with solvent molecules; as above, this calculation reproduces the observed intensities well in non-polar solvents, but poorly in polar ones.

5.2 Alkynes, HC≡CR

5.2.1 Crystal Structures

The crystal structures of several alkyne-containing compounds have been found to exhibit close, and nearly linear, contact between the H–C group and an oxygen atom of a neighbouring molecule. Thus, in compounds I [123], II [122] and III [60], the spatial relationship of the alkyne function and the carbonyl oxygen atom of another molecule is shown with each diagram. It is pertinent to keep in mind that these spatial factors are all derived from X-ray crystal structure analyses. Such analyses will give the relative positions of the

	I	II	III
$r(C \cdots O)$/pm	$321 \cdot 2 \pm 0 \cdot 9$	$326 \cdot 0 \pm 1 \cdot 5$	
$r(H \cdots O)$/pm	219	220	239
$\angle$ C–H$\cdots$O	$145 \cdot 3°$		$156°$

carbon and oxygen atoms very well; however, as Hamilton has shown [164], the position of the hydrogen atom is not located reliably using this technique. Similar close contacts involving alkyne groups and oxygen have been found by Calabrese *et al.* [61] in compounds IV and V:

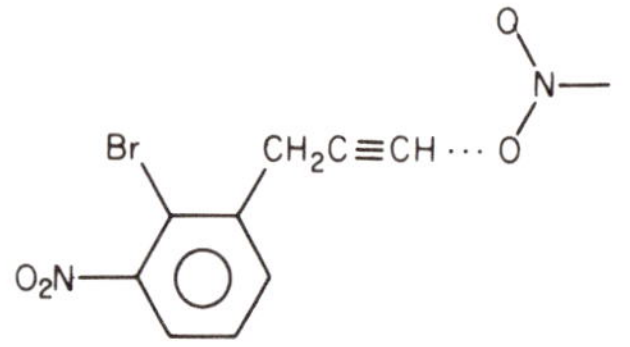

$r(\text{C}\ldots\text{O}) = 339$ pm
$r(\text{H}\ldots\text{O}) = 239$ pm
$\angle\,\text{C}-\text{H}\ldots\text{O} = 156°$

IV

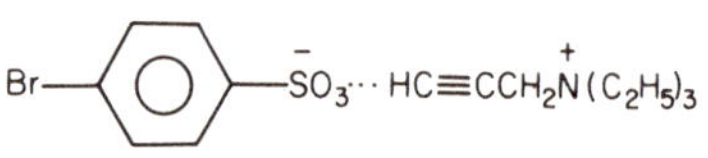

$r(\text{C}\ldots\text{O}) = 336$ pm
$r(\text{H}\ldots\text{O}) = 238$ pm
$\angle\,\text{C}-\text{H}\ldots\text{O} = 154°$

V

Although all the contacts cited here are at much less than the distances predicted from van der Waals radii, Calabrese *et al.* [61] assert that '. . . attempts to classify C–H···O contacts as hydrogen bonds solely on the basis of the H···O distance are unsound.'

5.2.2 Complex Formation Constants

A variety of formation constants for complexes of alkynes is given in Table 5.7 (see p. 135). The presence of electron-withdrawing groups in the acceptor molecule leads to a substantial increase in the value of K_ϕ, whereas alkyl groups appear to cause only a slight decrease.

Queignec and Wojtkowiak [316] showed that formation constants for alkyne complexes of diethyl ethanamide determined by means of gas–liquid chromatography agree very well with those measured using infrared spectroscopy. These authors note that the magnitude of K_ϕ is dependent upon the solvent used, falling into the order: tetrachloromethane < hexane < cyclohexane < 2,6,10,15,19,23-hexamethyltetracosane (squalane).

Goldstein, Mullins and Willis have reported [148] a very careful comparison of equilibrium quotients of formation for complexes of phenylethyne with a number of ethers and sulphides. The results obtained, after taking many precautions, from infrared spectroscopy and from NMR spectroscopy show significant discrepancies between the values obtained from the two techniques. They conclude that the differences must be due to the effects of solvation which were not allowed for in the models employed.

p. 124   *"Henry had a speculator's desire . . . ":* Mayer, *A Son of Thunder,* 467.

p. 125   *It did not stand that way for long:* on the architectural history of the Patrick Henry house, see James M. Elson, "A Short History of the Patrick Henry Memorial Foundation," *A Patrick Henry Essay,* (no. 11-94), The Patrick Henry Memorial Foundation, n.d.; and Mildred Stapley, "The Restoration of Red-Hill-on-the-Staunton," *Country Life in America* (Oct. 15, 1912), 37–40.

p. 129   *Against Interpretation:* Susan Sontag, *Against Interpretation and Other Essays* (New York: Dell Publishing Co., 1966).
          "Most serious thought in our time . . . ": Sontag, "The anthropologist as hero," in *Against Interpretation,* 69, 70.

pp. 130–131 *The Need for Roots:* Simone Weil, tr. Arthur Wills, preface by T. S. Eliot, *The Need for Roots: Prelude to a Declaration of Duties Toward Mankind* (New York: G. P. Putnam's Sons, 1952).

p. 131   *"A human being has roots . . . ":* Weil, *The Need for Roots,* 43.

Chapter Five

p. 136   *the Great Wagon Road:* see Parke Rouse, Jr., *The Great Wagon Road from Philadelphia to the South* (Richmond: The Dietz Press, 1995).

pp. 136 137 *August Spangenberg:* quoted in Rouse, *The Great Wagon Road,* 73, 75.

pp. 137–138 *The new settlers "lived like Savages . . . ":* this quote and citations following, from Charles Woodmason, Richard J. Hooker, ed., *The Carolina Backcountry on the Eve of the Revolution: The Journal and Other Writings of Charles Woodmason, Anglican Itinerant* (Chapel Hill: University of North Carolina Press, 1953), xxv, xxviii, xxx, 43, 45.

p. 143   *"Aeneas at Washington":* in William Pratt, ed., *The Fugitive Poets: Modern Southern Poetry in Perspective* (New York: E. P. Dutton & Co., 1965), 94–95.

p. 147   *The small treasury of inscriptions:* Margaret J. Watson and Louise M. Watson, *Tombstone Inscriptions from Family Graveyards in Greenwood County, S.C.* (Greenwood, S.C.: Drinkard Printing Co., 1972).

p. 148   *Walnut Grove Baptist Church:* Mrs. Willie Riley and Sarah Higgins, eds., *The History of Walnut Grove Baptist Church 1826-1976* (Greenwood, S. C.: Drinkard Printing Co., n.d.).

pp. 150–151 *Dr. Oliver Sacks's:* quotes from Oliver Sacks, *The Man Who Mistook His Wife for a Hat and Other Clinical Tales* (New York: Perennial Library, 1985), 109, 111.

p. 153   *the Federal Style*: see Jonathan Poston, "Federal and Empire (1780–1850)," in Stephen Calloway and Elizabeth Cromley, eds., *The Elements of Style: A Practical Encyclopedia of Interior Architectural Details from 1485 to the Present* (New York: Simon and Schuster, 1991), 204–231; and Spiro Kostof, *A History of Architecture: Settings and Rituals* (New York and Oxford: Oxford University Press, 1985), 617–618.

5.2.3 Enthalpies of Mixing

The enthalpies of mixing of several alkynes-1 with donor substances are given in Table 5.8 (see p. 137). A tendency is apparent toward lower values of $\Delta_m H$ as the size of the alkyne molecule increases. Kurkchi and Iogansen [229] suggest that it may be possible to extrapolate the observed values to obtain a value for the C≡C–H group with each donor. However, since $\Delta_m H$ combines the enthalpy of formation of each complex and its formation constant (see Section 3.1.3), even such extrapolated values are unlikely to be of much significance other than as a general indication of relative complex strengths.

5.2.4 Enthalpies of Complex Formation

The enthalpy change upon formation of an alkyne/donor complex is seen from the values in Table 5.9 (see p. 137) to be rather small, although comparable to the corresponding values for trichloromethane complexes (see Table 3.6).

5.2.5 Solubilities

The solubility of ethyne in a variety of solvents is given in Table 5.10 [85], together with the value calculated from Raoult's Law. The observed value is higher than expected in most of the solvents chosen; this is accounted for in terms of a hydrogen-bonding interaction between ethyne and solvent molecules. In the cases where self-association of either solute or solvent is likely to be more important (e.g. solvent = cyclohexane, ethanol, etc.), the observed solubility is smaller than predicted. It is interesting to note that (dimethylamino)benzene and nitrobenzene are not effective donor solvents toward ethyne.

Kiyama and Hiraoka [217] find much the same behaviour, in that the solubility of ethyne is greater than expected in oxolane, over a range of temperatures and pressures, but less than expected in benzene, methanol and water.

McKinnis [266] has devised a listing of relative donor abilities (towards ethyne) of various functional groups; these donor abilities are, then, directly related to the solubility of ethyne in solvents containing these functional groups. The assumption made in compiling such values (see Table 5.11, p. 139) is that solute–solvent hydrogen bonding is solely responsible for any difference in ethyne solubility from that expected on the basis of Raoult's Law.

5.2.6 Effects on the Vibrational Spectrum

The vibrational behaviour of a number of alkynes-1 upon complex formation has been studied extensively [16, 45, 50, 61, 64, 97, 120, 139, 140, 186, 189, 190, 191, 197, 275, 304, 354, 355, 363, 394, 402]. In particular, the thorough studies carried out at l'Université de Bordeaux have been summarised in very useful form in the doctoral thesis of Pham Van Huong [186], on the contents of which much of this section is based.

p. 155   *Composing a genealogical memoir . . . :* Samuel Edward Mays, *The Family of Mays,* (Plant City Florida, 1929), reproduced in Ivan K. Mays, *The Mays Family: A Sequel to "The Family of May," by Samuel Edward Mays* (n.p. [Austin, Texas], 1970).

Chapter Six

p. 169   *a film scholar:* Edward D. C. Campbell, Jr., in Wilson and Ferris, eds., *Encyclopedia of Southern Culture,* 923.

p. 186   *Jefferson's dream:* Thomas Jefferson, introduction by Thomas Perkins Abernathy, *Notes on the State of Virginia* (New York: Harper & Row, 1964).

p. 186   *". . . not only is it misguided . . . ":* Joseph J. Ellis, *American Sphinx: The Character of Thomas Jefferson* (New York: Alfred A. Knopf, 1997), 138, 250–251.

p. 188   *"We must now place the manufacturer . . . ":* Thomas Jefferson, quoted in Ellis, *American Sphinx,* 250.

p. 189   *"Though my young countrymen . . . ":* quoted in Clarence Poe, *True Tales of the South at War: How Soldiers Fought and Families Lived, 1861–1865* (Chapel Hill: The University of North Carolina Press, 1961), 128–129.

Chapter Seven

pp. 195–196 *Judith McGuire:* quoted in Poe, *True Tales,* 182.

p. 196   *. . . desertion from the reunited country . . . :* Isabel Vincent, *The Globe and Mail,* Saturday, 20 March 1993. See also Eugene C. Harter, *The Lost Colony of the Confederacy* (Jackson: The University Press of Mississippi, 1985).

p. 198   *. . . Ethel Cooper Turner:* quote from "History of Mayesville, South Carolina" (1983), manuscript in The Sumter County Historical Archives.

       Another of John's relations: details and quote from the war diaries of Samuel Elias Mays, supplied to me by Steven Griffin Mays.

p. 200   *"Redeemer Nation that died . . . ":* Charles Reagan Wilson, *Baptized in Blood: The Religion of the Lost Cause, 1885–1920* (Athens: The University of Georgia Press, 1980), 1.

p. 201   Book XXIII *of the Iliad:* tr. Richard Lattimore (Chicago: University of Chicago Press, 1951).

pp. 203–204 *C.S.S. Shenandoah:* the tale as told by William C. Whittle, "The Cruise of the Shenandoah," *Southern Historical Papers,* XXV (1907), 235–258.

p. 204   *Noble man! chivalrous soul!:* Capt. S. A. Ashe, "The Shenandoah," *Southern Historical Papers,* XXXII (1904), 322.

pp. 204–205 *the enormous ideological industry:* see Charles Reagan Wilson, *Baptized in Blood,* 18–19.

pp. 205–206 *The tidewater aristocrat General Lee:* quotes from Poe, *True Tales,* 204; and James M. McPherson, *Battle Cry of Freedom: The Civil War Era* (New York: Ballantine Books, 1988), 249–250.

p. 207   *"The true hero, the true subject . . . ":* Simone Weil, tr. Mary McCarthy, *The Iliad, or the Poem of Force* (Wallingford, Pa.: Pendle Hill, 1976), 3.

p. 208   *"The cries of the wounded Yankees . . . ":* Samuel Elias Mays, "Civil War Scenes," *Tyler's Quarterly Magazine,* reprinted in *Mays & Company,* No. 5 (Dec., 1991).

p. 211   *Unlike the young soldiers of Mayesville:* Ann Herd Bowen, *Greenwood County: A History* (Greenwood, S. C.: The Museum, 1992), 308–309.

p. 212ff.   *By the end of 1865:* history of east Texas from Leila B. LaGrone, *A History of Panola County, Texas 1819–1978* (Carthage, Texas: Panola County Historical Commission, 1979), 42ff.

Chapter Eight

pp. 225–236 *of photography:* Pierre Bourdieu, tr., Shaun Whiteside, *Photography: A Middle-brow Art* (Stanford: Stanford University Press, 1990); Roland Barthes, tr. Richard Howard, *Camera Lucida* (New York: Hill and Wang, 1981); Susan Sontag, *On Photography* (New York: Farrar, Straus and Giroux, 1977).

p. 237   *"The Lord who prophesies at Delphoi . . . ":* tr. Davenport, 7 *Greeks,* 160.

Chapter Nine

p. 268   *"the shrewdest thinkers and artists . . . ":* Susan Sontag, introduction to E. M. Cioran, tr. Richard Howard, *The Temptation to Exist* (New York: Quadrangle/The New York Times Book Co.), 8.

The $\equiv$C–H stretching vibration of alkynes-1 is found at about 3300 cm^{-1}, rendering its study convenient even in C–H-containing solvents; this is not the case with the other common class of C–H acceptor compounds, the trihalomethanes. The effects on the wavenumber of this stretching mode of various donor molecules are given in Table 5.12 (see pp. 139–151). Most of the wavenumber shifts are reported relative to the vapour-phase molecule; in cases where the original paper reported the shift with respect to a dissolved phase, the reported shift is given in parentheses, and the reference medium is in parentheses after the solvent.

In most donor solvents, two bands are apparent in the C–H stretching region of the spectrum, one arising from 'free' alkyne molecules, and the other from 'associated' molecules [186, 304]. Temperature and concentration studies establish unequivocally that the narrower, higher-wavenumber band is due to 'free' alkyne. In non-associating solvents, on the other hand, only one band is observed. The wavenumbers of the latter bands define a straight line Kirkwood–Bauer–Magat (KBM) graph [186, 191] when the relative shifts, $\Delta v/v$, are plotted against $(n^2 - 1)/(2n^2 + 1)$, where n is the refractive index of the solvent. Hence, these shifts are due to the polarisabilities of the solvents concerned. The 'free' molecule bands in donor solvents also fall on or near the line so defined.

The observed wavenumbers in solvents of the dihalomethane and trihalo-methane types define a line parallel to the KBM plot but displaced to higher wavenumber. This effect is postulated [186] to be due to a solute–solvent orientation effect not directly involving the C–H group.

The wavenumber shifts of associated alkynes are far in excess of the values predicted from the KBM graph. This excess shift is attributed to perturbation of the C–H bond by hydrogen bonding. If solvent groups are classified according to the magnitude of this excess shift, they fall into the following order: unsaturated hydrocarbons $\approx$ aromatic hydrocarbons $<$ nitriles $<$ ketones $\approx$ ethers $<$ azines $<$ tertiary amines $\approx$ sulphoxides.

The intensity and the linewidth of the C–H stretching mode are found to increase in more strongly donor solvents, as seen from Table 5.13 (see p. 151). This behaviour is considered to be characteristic of hydrogen bonding. The *relative* amplitudes of the 'free' and 'associated' C–H bands in a donor solvent are an indication of the extent of association of alkyne molecules; this is the criterion for assigning the two bands appropriately. Huong has shown [186] that the relative intensities of the two bands change monotonically in the series of homologues propanone, butanone, pentanone-3, heptanone-4 and nonanone-5, the proportion of associated molecules increasing as the number of oxygen atoms per unit volume increases; with the exception of propanone, the dependence is a near-linear one.

The stretching frequency of the triple bond in alkynes is virtually independent of environment, showing only a slight decrease, from the value in hexane, in highly polarisable solvents such as diiodomethane [186] (2113 cm^{-1} vs. 2121 cm^{-1}); only in tertiary amine solvents is there observed a separate band due to associated molecules of alkyne [141]. The intensity of this band, on the other hand, is found to decrease markedly as the solvent donor ability increases, and to increase as the solvent *acceptor* ability increases [186]. Then, the intensities decrease in the order of solvents: tribromomethane > tetrachloromethane > methylbenzene > nitromethane > heptanone-4 > triethylamine ≈ diether ether ≈ azine [141, 186]. Huong proposes [186] a qualitative explanation for this phenomenon in terms of the charge distribution on the triple bond, as

$$\overset{\delta+ \quad \delta-}{R-C\equiv C-H.}$$

Assuming a monotonic relationship between $\mu(C\equiv C)$ and $d\mu/dr$, he considers that a reduction of $\mu(C\equiv C)$ in the complex

$$R-C\equiv C-H\cdots X$$

or an enhancement thereof in the complex

$$\begin{array}{c} R \\ | \\ C \\ ||| \\ C\cdots H-X \\ | \\ H \end{array}$$

may account for the intensities observed. By way of confirmation, the $C\equiv C$ mode of ethyne is found to be inactive in the infrared spectrum in the solvents hexane and tetrachloromethane (as expected from symmetry considerations) but active in the acceptor solvents $HCCl_3$ and $HCBr_3$ as well as in donor solvents such as propanone and 1,3,5-trimethylbenzene [97]. This observation provides a neat proof of the presence of $\equiv C-H\cdots$ hydrogen bonding. This band, found at 1964 cm^{-1}, is forbidden in the free ethyne molecule but the partial removal of symmetry by complex formation permits it to appear. At high concentrations of propanone, the intensity of this band diminishes owing to the conversion of 1:1 complexes into 2:1 complexes in which, of course, the molecular symmetry is restored.

Alkyne molecules in which deuterium has been substituted for hydrogen-1 at the terminal position exhibit some interesting differences in behaviour from their sister compounds. Since the wavenumber shift $\Delta\nu/\nu$ for a diatomic vibrator is expected to be independent of the reduced mass of the vibrator, then, in the first approximation,

$$\left(\frac{\Delta\nu}{\nu}\right)_{XD} = \left(\frac{\Delta\nu}{\nu}\right)_{XH} \tag{5.1}$$

However, for the groups $\equiv CH$ and $\equiv CD$, just as is the case for $HC\equiv N$ and $DC\equiv N$, it is found that

$$\left(\frac{\Delta\nu}{\nu}\right)_{XD} < \left(\frac{\Delta\nu}{\nu}\right)_{XH} \tag{5.2}$$

Rather, the postulate is made of mechanical coupling between the ≡CH (or CD) and C≡C stretching modes; then, the relation becomes

$$\left(\frac{\Delta\nu}{\nu}\right)_{CH} + \left(\frac{\Delta\nu}{\nu}\right)_{C\equiv C(H)} = \left(\frac{\Delta\nu}{\nu}\right)_{CD} + \left(\frac{\Delta\nu}{\nu}\right)_{C\equiv C(D)} \qquad (5.3)$$

The values of these four parameters for heptyne-1 in a range of solvents are given in Table 5.12. It is seen that, while $(\Delta\nu/\nu)_{CD} < (\Delta\nu/\nu)_{CH}$ and $(\Delta\nu/\nu)_{C\equiv C(D)} > (\Delta\nu/\nu)_{C\equiv C(H)}$, nevertheless the equality shown as equation 5.3 is a valid one.

Whereas $\nu_{C\equiv C(H)}$ was found to be virtually unaffected by a change of solvent, $\nu_{C\equiv C(D)}$ is much more strongly affected, to the extent that separate bands due to 'free' and 'associated' molecules of heptyne-1 are observed; two bands are observed for $\nu_{C\equiv C(H)}$ of phenylethyne only in triethylamine solvent [139]. This observation suggests that the coupling is strong between the C–D and C≡C(D) vibrators but relatively weak between C–H and C≡C(H).

In the case of a diatomic vibrator, HX, the expression (A/ν^2), where A is the intensity of the band at wavenumber ν, is expected to be independent of the reduced mass; but once again it is found that

$$\left(\frac{A}{\nu^2}\right)_{CD} < \left(\frac{A}{\nu^2}\right)_{CH} \qquad (5.4)$$

for alkynes [186], as for HCN [63]. However, for the case of mechanical coupling in a polyatomic molecule, it is expected that

$$\left(\frac{A}{\nu^2}\right)_{CH} + \left(\frac{A}{\nu^2}\right)_{C\equiv C(H)} = \left(\frac{A}{\nu^2}\right)_{CD} + \left(\frac{A}{\nu^2}\right)_{C\equiv C(D)} \qquad (5.5)$$

As may be seen from the data in Table 5.13, even this relationship is not obeyed, leading to the speculation that coupling with still further molecular vibrational modes must be responsible; yet some other factor must be operative as well since the relationship is not exactly obeyed by HCN and DCN [186].

Boobyer has pointed out [45] that the anharmonicity constant for the ≡C–H bond of pentyne-1 shows a marked decrease in strongly donor solvents (see Table 5.16, p. 152) and draws the conclusion that, since the vibrational energy levels become more uniformly spaced as complex formation becomes more complete, the potential energy function for the ≡C–H bond resembles more closely that of an harmonic oscillator when in a complex than when in a 'free' molecule.

An attempt to calculate the intensities of the ≡C–H vibrational mode of pentyne-1 in various solvents on the basis of a simple electrostatic model is moderately successful [45]. Boobyer asserts that such a model gives more satisfactory predictions than do the bulk dielectric theories.

Acknowledgments

POWER IN THE BLOOD IS NEITHER HISTORY NOR GENEALOGY, THOUGH THE text is grounded, necessarily, in historical and genealogical fact. For his help in steadying my speculations on the firmest possible foundations, I am especially indebted to Joseph Barron Chandler, Jr., my North Carolina cousin, a rigorous genealogist, and new friend, and to my Texas cousin Steven Griffin Mays, who generously aided my research and thought until his untimely death in the summer of 1996. The compendious gatherings of scattered genealogical evidences by Selena M. Dulac also saved me much laborious archival work.

I am grateful for the support given me thoughout this writing by my editors, Larry Ashmead and Jason Kaufman at HarperCollins in New York, and Jackie Kaiser at Penguin Books of Canada in Toronto; Jan Whitford, my literary agent, and Westwood Creative Artists; Anne Collins and Antanas Sileika, friends and excellent critics who, along with Joe Chandler, read and generously commented on early versions of this text; and Eleanor Mikucki, who copy-edited the completed manuscript.

Among the many other witnesses who aided me in a variety of ways are Marjorie Mays Arthur; Essie De Harris; John Mays Bateman, Sr, and Melissa Mayfield Bateman; Mrs. Malcolm Earle; May Whitten; Lisa Speers; Richard Howorth; Sadie M. Starns Hoke; Lynda Mayes Dabbs and Billy Dabbs; James Elson; Elaine Martin; Rutledge Dingle; Daren Fonda; Lola Smith; Dr. Samuel H. deMent; Laurie K. Robinson; Barbara Pate; the Reverend Rodney L. Caulkins; Eugene Davis; Edith C. Poindexter; Isabel Vincent; the Reverend Harry R.

Mays; Jack McIvor; Mary Wark; the Reverend Debra Griffis-Wood-berry; William Rueckert; Howard Bahr; Opal Worthy; Lois C. Mar-bert; Anne Bentley Mays DuChateau, my sister; and the sort of friend most writers need when doing books: Robert Tugwell, my bank manager. Cathrin Bradbury, features editor of *The Globe and Mail,* generously allowed me a leave of absence to finish the writing. Whatever is sturdy in the architecture of this book is due to the con-tributions of these relatives, friends, colleagues, helpers; I alone bear responsibility for what is frail or false.

The phrase "geography of imagination," and all that this touch-stone opened to me, I owe to the American writer Guy Davenport.

My special thanks go to our family friend Helen Heller, the first person to tell me I should write about the South.

Laurence and Wojtkowiak [237] claim to have found a linear relationship of the Hammett type [166]:

$$\Delta\nu_{CH} - \Delta\nu_{CH}^0 = \rho\sigma \qquad (5.6)$$

between the gas-to-solution frequency shifts of the terminal proton of phenylethyne in various azines and the Hammett substituent constants (with 6 azines, the correlation coefficient $r = 0.990$). They mention that $\Delta\nu_{CH}^0 = 101 \pm 10$ cm^{-1} for azine itself, while $\rho = -68$ cm^{-1} for *meta* and *para* substituted azines, but $\rho = -118$ cm^{-1} for *ortho* substituted azines. Since no experimental data are given, these claims must be regarded as doubtful until verified.

It has been found that the $\equiv$C–H deformation mode shifts to higher frequency upon formation of a complex with diethyl ether [50]. The band due to associated alkyne molecules has 2 to 3 times the intensity of that due to unassociated molecules. Table 5.17 (see p. 153) contains the relevant wavenumbers.

Taken in conjunction with the $\equiv$C–H stretching mode behaviour, these observations show that association of alkynes with donor molecules results in vibrational behaviour much more characteristic of ordinary hydrogen bonds than does association of trihalomethane molecules.

5.2.7 Effects on the NMR Spectrum

The proton NMR solvent shifts of alkynes-1 are collected in Table 5.18 (see p. 153). Their chemical shift behaviour upon dilution in non-associating solvents such as tetrachloromethane or cyclohexane has been interpreted in terms of self-association of the alkyne molecules [34, 114, 171, 302, 408]. Alkynes-1 with an oxygen atom exhibit larger dilution shifts than those without [34], suggesting that the oxygen atom acts as the donor site in preference to the electrons of the triple bond. In addition, an aromatic ring seems to predominate over the triple bond in a molecule such as phenylethyne, in which the dilution shift is to higher field [408] whereas that of heptyne-1 is to lower field [34]. However, this conclusion is uncertain, since a downfield dilution shift of phenylethyne has also been reported [34].

The solvent shifts are, in general, proportional to the donor ability of the solvent molecule. A curious result is that of Braillon for hexyne-1 dissolved in benzene, whose solvent shift he reports as -0.087 ppm [49] whereas other workers obtain a value of $+0.03$ ppm [171]. This shift is expected by analogy with trihalomethanes and with other alkynes-1 to be positive; the reason for the discrepancy is not clear. Hatton and Richards suggested [171] that alkynes-1 do not form complexes with aromatic solvent molecules, since their solvent shifts with respect to cyclohexane are small; other workers have since shown that this speculation is valid [82].

The solvent shifts of the ^{13}C nucleus in $C_6H_5C\equiv^{13}CH$ are approximately three times as great as those of the terminal hydrogen, as may be seen from the data of Table 5.19 [408] (see p. 157). The exceptions are those solvents containing a benzene ring, in which the proton is shifted to high field–because of preferred orientations of the hydrogen atom with respect to the aromatic ring of the solvent –but the ^{13}C shift is to lower field. Any complex geometry which places the carbon atom away from the axis of the ring will result in a smaller shielding from the anisotropy effect. Alternatively, ^{13}C chemical shifts are probably dominated by factors other than the anisotropy of the solvent molecules.

Coleman, Sataty and Tyrrell [82] have reported the results of a detailed statistical analysis of the proton NMR spectra of 3-chloropropyne-1, 3-bromopropyne-1 and 1,4-dichlorobutyne-2 in a series of mixtures of cyclohexane and benzene solvents. These indicate the formation of 1:1 complexes of all three compounds with benzene, and the probability of a 1:2 complex of the dichloro compound. The relevant parameters obtained are given in Table 5.20 (see p. 157).

A very interesting result of this study is the observation that the monomer-to-dimer chemical shift change is twice as large for the methylene protons as for the ethynyl proton*, and also that the methylene shift difference in the dichloro compound is very similar to those in the 3-substituted propynes. By analogy with compounds of the H_2CX_2 type, the complex with benzene appears to have the geometry shown. The possibility of the existence of two separate complexes, one involving interaction of the $\equiv$CH group with benzene, is definitely excluded by the analysis of the data.

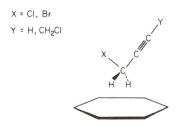

X = Cl, Br
Y = H, CH₂Cl

This result provides a puzzling contrast with the infrared observations of 3-bromopropyne-1 [7]; in CCl_4 solutions containing a strong donor molecule (dimethyl sulphoxide or azine), the terminal $\equiv$C–H group exhibits a frequency shift of more than 100 cm^{-1}, whereas the CH_2 modes are reported to be unaffected. Whether this paradox indicates that aromatic donors interact preferentially with a CH_2 group while σ donors interact with $\equiv$CH would appear to warrant further investigation.

The ethynyl proton shift of phenylethyne in cyclohexane solutions of tris(dimethylamino)phosphine oxide, and also the terminal ethynyl ^{13}C shift ($-C\equiv^{13}C-H$) have been measured for various ratios of concentrations of the alkyne and the donor species [408]. The formation constant for the 1:1 complex is listed in Table 5.7 at three temperatures; the values from ^{1}H and ^{13}C NMR

* This observation contrasts sharply with that of Hatton and Richards [171] that the chemical shift changes of CH₂ and ≡CH are approximately equal in aromatic solvents.

studies were found to be the same. The chemical shift differences between 'free' and 'associated' phenylethyne are given here:

T/K	273	295·2	314·5
Δ_c/ppm [^{1}H]	2·19 ± 0·02	2·10 ± 0·02	2·02 ± 0·02
Δ_c/ppm [^{13}C]		6·12 ± 0·04	6·20 ± 0·04

The ^{13}C shifts are self-consistent at the two temperatures considered; the proton shifts, on the other hand, show a tendency to decrease as the temperature is raised. Muller and Reiter [273] have predicted such behaviour for hydrogen bonded complexes on the basis of vibrational excitation changes with temperature; their model predicts a dissociation energy of approximately 25 kJ mol^{-1} for a gradient of 0·0039 ppm K^{-1}. Presumably, the ^{13}C shift is insensitive to changes in vibrational state.

As mentioned in Section 3.1.6.2, the magnitude of the ^{13}C–^{1}H coupling constant of trichloromethane is rather sensitive to changes of solvent, presumably because in donor solvents, a high proportion of HCCl$_3$ molecules is associated with solvent molecules [116]. However, although phenylethyne is known to form complexes with the same solvents, the coupling constant of the terminal C–H bond is much less sensitive to a change of solvent [116]. Presumably, this observation results from some difference in the nature of the two complexes.

Solvent	$^1J_{CH}$/Hz	
	HCCl$_3$	HC≡CC$_6$H$_5$
CCl$_4$	208·4	250·8
dimethyl sulphoxide	217·7	252·0
Difference	9·3	1·2

TABLE 5.1

Relative permittivities, $\epsilon_r (=\epsilon/\epsilon_0)$, of liquid HCN [80]
$\epsilon_0 = 8·85 \times 10^{-12}$ C^2 N^{-1} m^{-2}

T/K	ϵ_r	
258·2	213·2	supercooled
259·5	205·5	freezing point
263·2	191·9	
268·2	173·7	
273·2	158·1	(158·45 [223])
278·2	144·8	
283·2	133·1	
288·2	123·5	
291·2	118·3	
293·2	114·9	
298·2	106·8	
298·9	105·7	boiling point

TABLE 5.2

Formation constants for complexes of HCN

Donor	T/K	K	Units	Ref.
NH_3	–	2.7×10^{-5}	Pa^{-1}	[194]
diethyl ether	–	$(0.13 \pm 0.02) \times 10^{-5}$	Pa^{-1}	[344]
$2HCN \rightleftharpoons (HCN)_2$	273.1	0.094×10^{-5}	Pa^{-1}	[143]
	303.0	0.055×10^{-5}	Pa^{-1}	[143]
	333.1	0.034×10^{-5}	Pa^{-1}	[143]
	373.1	0.020×10^{-5}	Pa^{-1}	[143]
	462.3	0.010×10^{-5}	Pa^{-1}	[143]
$3HCN \rightleftharpoons (HCN)_3$	273.1	0.054×10^{-10}	Pa^{-2}	[143]
	303.0	0.0085×10^{-10}	Pa^{-2}	[143]
	333.1	0.0029×10^{-10}	Pa^{-2}	[143]
	373.1	0.0013×10^{-10}	Pa^{-2}	[143]

TABLE 5.3

The vibrational spectrum of HCN in various phases

	ν_1/cm^{-1} C≡N	ν_2/cm^{-1} bend	ν_3/cm^{-1} C–H	Ref.
	Vapour phase			
[1]HCN (monomer)		712	3312	[176]
		712		[2]
	2096.7	712.1	3311.4	[305]
			3312	[344]
	2096.68	712.26	3311.47	[319]
	2089	712	3312	[371]
[1]H[13]CN (monomer)	2062.268	707.393	3293.506	[319]
[1]HCN (dimer)	2095			[194]
[2]HCN (monomer)	1921	569	2629	[371]
	Liquid phase			
[1]HCN (292 K)	2096	ca. 794	3215	[305]
(261 K)	2097	ca. 798	3207	[305]
[2]HCN (292 K)	1909	ca. 615	2581	[305]
(261 K)	1908	ca. 622	2575	[305]
	Solid phase			
[1]HCN	2097	828, 838	3135	[344]
(78 K)	2098	819	3145	[305]
(96 K)	2097	828, 838	3132	[176]
(183 K)	2098	827, (840)	3129	[305]
[2]HCN	1888	650	2550	[344]
(78 K)	1890	643	2553	[305]
(183 K)	1885	648, 653	2545	[305]
	Matrix isolation at 20.5 K			
[1]HCN in A (monomer)	2093.4	720.2	3303.3	[216]
in N_2 (monomer)	2097.3	745.6, 736.0	3287.6	[216]
in CO (monomer)	2104	761.3, 739.0	3261.2	[216]
in A (dimer)	2114.4, 2090.4	797.3, 732.4	3301.3, 3202.0	[216]
in N_2 (dimer)	2110.9, 2092.9	799.0, 757.6	3282.0, 3204.9	[216]
in CO (dimer)	2112, 2088	803.0, 778.6	3244.4, 3201.3	[216]

[2]HCN in A (monomer)	1922·7	572·0	2631·3	[216]
in N$_2$ (monomer)	1920·6	594·0, 588·4	2617·8	[216]
in A (dimer)	1935·4, 1900·0	622·0, 579·5	2626·7, 2573·9	[216]
in N$_2$ (dimer)	1927·0, 1898·4	628·5, 603·0	2616·0, 2574·4	[216]

TABLE 5.4

The vibrational spectrum of HCN in complexes

Donor	ν_1/cm^{-1}	ν_2/cm^{-1}	ν_3/cm^{-1}	Ref.
		[1]HCN		
vapour (unassociated)	2096·68	712·26	3311·47	[319]
dimethyl ether (vapour)	2085		3185 ± 2	[344]
(solid)	2075	860	3005	[344]
ammonia (vapour)	2086·2		3150	[194]
methyleneimine (matrix)	2076	862·5	3018	[269]
		[2]HCN		
vapour	1928·0	569·1	2630·3	[305]
dimethyl ether (vapour)	1900		2565 ± 15	[344]
(solid)	1838	666, 683	2490	[344]
ammonia (vapour)	1890		2559	[194]
methyleneimine (matrix)	1844	670·5	2495	[269]

TABLE 5.5

Solvent dependence of the stretching frequencies of HCN [64]

Solvent	$10^4 \, \Delta\nu_s/\nu_s$					
	HCN			DCN		
	HC	CN	Total	DC	CN	Total
hexane	100					
tetrachloromethane	103	5		76		
trichlorobromomethane	118	26				
carbon disulphide	146			106		
trichloromethane	125	10	135	87	44	131
pentachloroethane	130	17	147	91	44	135
tribromomethane	156	31	187	112	65	177
dichloromethane	157	17	174	108	60	168
dibromomethane	178	31	209	125	73	198
diiodomethane	217	45				
1,1,1-trichloroethane	133	26	159	95	47	142
1,1-dichloroethane	163	26				
1,2-dichloroethane	190	26	216	129	73	202
1,2-dichloropropane	192			129	73	
1,2-dibromoethane	199	36	235	137	80	217
1,2-dibromopropane	208			137	80	
2,3-dibromobutane	217	36				
2-methyl-1-chloropropane	202	29	231	137	78	215
2-methyl-2-chloropropane	208	29				

TABLE 5.5—*continued*

Solvent	$10^4 \, \Delta\nu_s/\nu_s$					
	HCN			DCN		
	HC	CN	Total	DC	CN	Total
2-chlorobutane	210	29	*239*	141	78	*219*
1-bromopropane	217	36				
2-methyl-2-bromopropane	227	36				
iodomethane	227	43				
2-methyl-2-iodopropane	237					
trichloroethanenitrile	217	26	*243*	146	80	*226*
chloroethanenitrile				228	101	
ethanenitrile	347	33	*380*	236	138	*374*
nitromethane	239	24	*263*	162	86	*248*
nitroethane	242	33	*275*	171	93	*264*
1-nitropropane	252	29	*281*	171	91	*262*
2-nitropropane	272	31	*303*	175	99	*274*
propanone		57		291	179	
1,2,4-trichlorobenzene	151	26				
1,2-dichlorobenzene	160	31				
chlorobenzene	175	33	*208*	127	73	*200*
4-chloro-1-methylbenzene	190	36				
bromobenzene	190	36				
3-chloro-1-methylbenzene	192	33				
2-chloro-1-methylbenzene	192	33				
benzene	239	43	*282*	156	96	*252*
methylbenzene	263	45	*308*	171	104	*275*
1,2-dimethylbenzene	284	52				
1,3-dimethylbenzene	285	48	*333*	188	112	*300*
1,4-dimethylbenzene	290					
1,3,5-trimethylbenzene	313					
(dimethylamino)benzene	338			220	138	

TABLE 5.6

Ratios of band intensities of HCN in solution and in the vapour [63]

Solvent	A_s/A_v			
	H–CN	D–CN	HC≡N	DC≡N
tetrachloromethane	1·8		19	
	1·4[a]		15[a]	
carbon disulphide	1·4	1·1		
trichlorobromomethane	2·0			
trichloromethane	1·9	1·5	32	4·9
	1·5[a]	1·2[a]	25[a]	3·8[a]
dichloromethane	2·5	1·9	36	5·9
tribromomethane	2·1	1·7	39	5·0
	1·5[a]	1·2[a]	26[a]	3·4[a]
dibromomethane	2·3		40	
1,2-dichloroethane	2·6	1·9	42	(7·8)
2-(chloromethyl)propane			42	
1,2-dibromoethane	2·7	1·9	44	5·7

2-(bromomethyl) propane				57	7·0
trichloroethanenitrile	3·4	2·2			8·1
ethanenitrile				128	
				107[a]	
nitromethane				87	
nitroethane	3·9	2·5		87	11
	3·2[a]	2·1[a]		71[a]	9·0[a]
propanone				164	
1,2,4-trichlorobenzene	2·2			31	
chlorobenzene	2·5			33	
benzene	2·6			50	
	1·9[a]			38[a]	
methylbenzene	2·6			54	
1,3-dimethylbenzene	3·1			50	

[a] Ref. [144].

TABLE 5.7

Formation constants for $RC{\equiv}CH$ complexes

R	Donor		T/K	K_ϕ	Units	Solvent	Ref.
H	ammonia		323·2	0·00729	atm^{-1}	(vapour)	[72]
			348·2	0·00579	atm^{-1}	(vapour)	[72]
			373·2	0·00473	atm^{-1}	(vapour)	[72]
			398·2	0·00392	atm^{-1}	(vapour)	[72]
			423·2	0·00322	atm^{-1}	(vapour)	[72]
	propanone		298	0·36	M^{-1}	CCl$_4$	[97]
		(1:2)	298	0·045	M^{-1}	CCl$_4$	[97]
C_5H_{11}	diethyl ethanamide		301	0·60 ± 0·02	M^{-1}	(vapour)	[316]
			298	0·60 ± 0·03	M^{-1}	squalane	[316]
			301	0·60 ± 0·02	M^{-1}	CCl$_4$	[316]
	diethyl ether		298	1·2 ± 0·2	mf^{-1}	CCl$_4$	[186]
	ethanenitrile		298	2·1 ± 0·3	mf^{-1}	CCl$_4$	[186]
	1,4-dioxane		298	3·5 ± 0·4	mf^{-1}	CCl$_4$	[186]
	propanone		298	3·2 ± 0·4	mf^{-1}	CCl$_4$	[186]
			298	0·28 ± 0·04	M^{-1}	CCl$_2$=CCl$_2$	[54]
			298	0·28 ± 0·04	M^{-1}	CS$_2$	[54]
			298	0·29 ± 0·04	M^{-1}	CF$_2$ClCFCl$_2$	[54]
			298	0·32 ± 0·04	M^{-1}	CCl$_4$	[54]
			298	0·43 ± 0·04	M^{-1}	C$_{10}$H$_{22}$	[54]
			298	0·43 ± 0·04	M^{-1}	C$_{14}$H$_{30}$	[54]
			298	0·44 ± 0·05	M^{-1}	c-C$_6$H$_{12}$	[54]
			298	0·45 ± 0·04	M^{-1}	C$_6$H$_{14}$	[54]
$(CH_3)_3C$	diethyl ethanamide		273	0·96 ± 0·04	M^{-1}	squalane	[316]
			298	0·58 ± 0·03	M^{-1}	squalane	[316]
			301	0·60	M^{-1}	squalane	[316]
			301	0·52	M^{-1}	c-C$_6$H$_{12}$	[316]
			301	0·46	M^{-1}	C$_7$H$_{16}$	[316]
			301	0·35	M^{-1}	CCl$_4$	[316]

[a] Measured by ^{1}H NMR spectroscopy.
[b] Measured by infrared spectroscopy.
[c] Measured by ^{13}C NMR spectroscopy.

TABLE 5.7–*continued*

R	Donor	T/K	K_ϕ	Units	Solvent	Ref.
$(CH_3)_3Si$	diethyl					
	ethanamide	301	0.65 ± 0.02	M^{-1}	CCl_4	[316]
		298	0.66 ± 0.03	M^{-1}	squalane	[316]
		273	1.12 ± 0.04	M^{-1}	squalane	[316]
$(C_2H_5)_3Si$	diethyl					
	ethanamide	301	0.65 ± 0.02	M^{-1}	CCl_4	[316]
C_6H_5	diethyl ether	302	1.1	mf^{-1}	CCl_4	[50]
		306.7	0.43^a	mf^{-1}	CCl_4	[148]
		306.7	0.32^b	mf^{-1}	CCl_4	[148]
	ethyl 2-chloro-					
	ethyl ether	306.7	0.39^a	mf^{-1}	CCl_4	[148]
		306.7	0.78^b	mf^{-1}	CCl_4	[148]
	bis(2-chloro-					
	ethyl) ether	306.7	0.55^a	mf^{-1}	CCl_4	[148]
		306.7	0.82^b	mf^{-1}	CCl_4	[148]
	di(2-propyl) ether	306.7	0.69^a	mf^{-1}	CCl_4	[148]
		306.7	0.66^b	mf^{-1}	CCl_4	[148]
	dibutyl ether	306.7	0.60^a	mf^{-1}	CCl_4	[148]
		306.7	0.93^b	mf^{-1}	CCl_4	[148]
	dihexyl ether	306.7	0.51^a	mf^{-1}	CCl_4	[148]
		306.7	2.19^b	mf^{-1}	CCl_4	[148]
	oxolane	306.7	0.21^a	mf^{-1}	CCl_4	[148]
		306.7	0.48^b	mf^{-1}	CCl_4	[148]
	2-methyloxolane	306.7	0.27^a	mf^{-1}	CCl_4	[148]
		306.7	0.48^b	mf^{-1}	CCl_4	[148]
	oxane	306.7	0.23^a	mf^{-1}	CCl_4	[148]
		306.7	0.71^b	mf^{-1}	CCl_4	[148]
	diethyl sulphide	306.7	0.07^a	mf^{-1}	CCl_4	[148]
		306.7	0.10^b	mf^{-1}	CCl_4	[148]
	thiolane	306.7	0.06^a	mf^{-1}	CCl_4	[148]
		306.7	0.20^b	mf^{-1}	CCl_4	[148]
	azine		$0.1–0.2$	M^{-1}	CCl_4	[225]
	tris(dimethyl-					
	amino) phosphine					
	oxide	273.0	18.1 ± 0.5	mf^{-1}	$c\text{-}C_6H_{12}$	[408]
		295.2	11.3 ± 0.3^a	mf^{-1}	$c\text{-}C_6H_{12}$	[408]
		295.2	10.7 ± 0.3^c	mf^{-1}	$c\text{-}C_6H_{12}$	[408]
		314.5	7.4 ± 0.2^a	mf^{-1}	$c\text{-}C_6H_{12}$	[408]
		314.5	7.5 ± 0.2^c	mf^{-1}	$c\text{-}C_6H_{12}$	[408]
C_6H_5CO	diethyl ether	302	2.0 ± 0.2	mf^{-1}	CCl_4	[50]
$(CH_3)_2ClC$	diethyl					
	ethanamide	301	0.87 ± 0.02	M^{-1}	CCl_4	[316]
		273	1.48 ± 0.06	M^{-1}	squalane	[316]
		298	0.90 ± 0.04	M^{-1}	squalane	[316]
$ClCH_2$	diethyl					
	ethanamide	301	1.54 ± 0.02	M^{-1}	CCl_4	[316]
		273	2.78 ± 0.10	M^{-1}	squalane	[316]
		298	1.73 ± 0.07	M^{-1}	squalane	[316]
$BrCH_2$	diethyl					
	ethanamide	273	2.74 ± 0.10	M^{-1}	squalane	[316]
		298	1.67 ± 0.07	M^{-1}	squalane	[316]
		301	1.50	M^{-1}	squalane	[316]
		301	1.17	M^{-1}	$c\text{-}C_6H_{12}$	[316]
		301	0.90	M^{-1}	C_7H_{16}	[316]
		301	0.60	M^{-1}	CCl_4	[316]

TABLE 5.8

Enthalpies of mixing of alkynes-1 [mole fraction = 0·5]

Donor	T/K	$\dfrac{\Delta_m H}{kJ\ mol^{-1}}$	Ref.
$C_6H_5C{\equiv}CH$			
methyl ethanamide	276	−0·710	[85]
aminocyclohexane	276	−0·755	[85]
propanone	276	−0·795	[85]
diethyl ether	276	−1·130	[85]
dimethyl ethanamide	276	−2·720	[85]
$C_3H_7C{\equiv}CH$			
tribromomethane	−	−3·510	[229]
dimethyl methanamide	−	−4·900	[229]
$C_2H_5C{\equiv}CH$			
tribromomethane	−	−4·10	[229]
1,4-dioxane	−	−3·26	[229]
methyl ethanoate	−	−4·39	[229]
propanone	−	−5·31	[229]
cyclohexanone	−	−6·61	[229]
dimethyl methanamide	−	−6·15	[229]
1-methylazolidinone-2	−	−7·82	[229]
tris(dimethylamino)phosphine oxide	−	−8·03	[229]
$CH_3C{\equiv}CH$			
tribromomethane	−	−4·27	[229]
1,4-dioxane	−	−3·97	[229]
methyl ethanoate	−	−5·31	[229]
propanone	−	−6·32	[229]
cyclohexanone	−	−7·74	[229]
dimethyl methanamide	−	−7·41	[229]
1-methylazolidinone-2	−	−9·29	[229]
tris(dimethylamino)phosphine oxide	−	−9·79	[229]
$HC{\equiv}CH$			
tribromomethane	−	−1·21	[229]
1,4-dioxane	−	−4·64	[229]
methyl ethanoate	−	−4·98	[229]
propanone	−	−6·19	[229]
cyclohexanone	−	−7·03	[229]
dimethyl methanamide	−	−7·78	[229]
1-methylazolidinone-2	−	−9·20	[229]
tris(dimethylamino)phosphine oxide	−	−10·88	[229]

TABLE 5.9

Enthalpies of formation of $RC{\equiv}CH$ complexes

R	Donor	T/K	$\dfrac{-\Delta_\phi H}{kJ\ mol^{-1}}$	$\dfrac{-\Delta_\phi S}{J\ K^{-1}\ mol^{-1}}$	Ref.
$(CH_3)_3C$	azine		5·0 ± 2·1		[142]
	diethyl ethanamide	301	13·8 ± 2·5		[316]
C_5H_{11}	1,4-dioxane	298	4·6	10·9	[186]
	propanone	298	4·6	11·2	[186]

TABLE 5.9–*continued*

R	Donor	T/K	$\dfrac{-\Delta_\phi H}{\text{kJ mol}^{-1}}$	$\dfrac{-\Delta_\phi S}{\text{J K}^{-1}\text{ mol}^{-1}}$	Ref.
C_5H_{11}–*cont.*	ethanenitrile	298	5·4	15·5	[186]
	diethyl ether	298	5·9 ± 2·1	19·1	[186]
$(CH_3)_3Si$	diethyl ethanamide	301	14·2 ± 2·5		[316]
$C_6H_5(CH_3)_2Si$	azine		11·5 ± 1·9		[142]
$(CH_3)_2Si{<}$	azine		10·5 ± 1·5		[142]
$(CH_3)_3SiC{\equiv}C$					
$(CH_3)_2Si{\diagup}$	azine		15·0 ± 2·5		[142]
$(C_2H_5)_3Sn$	azine	302	8·4 ± 1·3		[142]
C_6H_5	diethyl ether	302	5·9 ± 1·3	18·8	[50]
	tris(dimethylamino)-phosphine oxide	273·0–314·5	15·5 ± 1·3		[408]
$BrCH_2$	diethyl ethanamide		12·6 ± 2·1		[316]
$ClCH_2$	diethyl ethanamide		12·6 ± 2·1		[316]
$Cl(CH_3)_2C$	diethyl ethanamide		13·4 ± 2·1		[316]

TABLE 5.10

Solubilities of ethyne [85]

Solvent	T/K	Solubility/mol mol^{-1}	
		measured	calculated
2-ethoxy-3-oxapentane	263	0·200	0·04
2,4-dioxapentane	263	0·192	0·04
ethanal	263	0·177	0·04
ethyl ethanoate	263	0·168	0·04
ethyl methanoate	263	0·144	0·04
methyl ethanoate	263	0·163	0·04
methyl methanoate	263	0·142	0·04
3-methylbutyl ethanoate	263	0·166	0·04
3-methylbutyl methanoate	263	0·095	0·04
ethanoic acid	291	0·0150	0·0245
ethanol	291	0·0151	0·0245
benzene	277	0·0252	0·0342
aminobenzene	277	0·0223	0·034
(dimethylamino)benzene	273·5	0·037	0·0382
nitrobenzene	277	0·0223	0·0345
cyclohexane	276	0·0175	0·0352
propanone	273·15	0·092	0·038
	288	0·074	0·026
	298	0·0402	0·0206
[12 atm] 288		0·490	0·312

TABLE 5.11

Relative donor abilities of functional groups toward ethyne as the acceptor molecule [266]

Donor group	Relative donor ability	Donor group	Relative donor ability
$>P \to O$	7·66	$>C-Br$	2·09
$>P-O-P<$	7·64	$>C-N<$	1·59
$>S \to O$	4·97	$-C\equiv N$	1·32
$>C-F$	4·21	$>N=O$	1·16
$>C-O<$	2·93	$>C-S<$	0
$>C-Cl$	2·73	$>C-I$	0
$>C=O$	2·70	$>C-H$	0

Reprinted from *Ind. and Eng. Chemistry*, Vol. 47, No. 4, 1969, p. 851. Copyright 1969 by the American Chemical Society. Reprinted by permission of the copyright owner.

TABLE 5.12

Vibrational frequency changes, linewidths, and intensities of the $\equiv C-H$ stretching mode of alkynes-1 upon complex formation

Donor	$10^4\ \Delta\nu/\nu$	$w_{1/2}/cm^{-1}$	Intensity	Solvent	Ref.
	HC≡CH				
octane	15			donor	[275]
2,2,4-trimethylpentane	73			donor	[189], [190]
cyclohexane	73		$1·5^a$	donor	[189], [190]
tetrachloromethane	79		$1·9^a$	donor	[189], [190]
	61			donor	[140]
	67			donor	[191]
carbon disulphide	103		$1·7^a$	donor	[189], [190]
	82			donor	[191]
ethyne[liquid]	95			pure liquid	[354]
trichloromethane	85			donor	[191]
	70			donor	[140]
tribromomethane	116			donor	[189], [190]

^{a} Ratio of intensities in solution and in the vapour: A_S/A_V.
^{b} Units: $dm^3\ mol^{-1}\ cm^{-2}$.
^{c} Units: $dm^3\ mol^{-1}\ cm^{-2}$.
^{d} Units: $m^3\ mol^{-1}\ 10^{-7}$.